COSMOLOGY AND
PARTICLE PHYSICS

Proceedings of the Theoretical Workshop on

COSMOLOGY AND PARTICLE PHYSICS

July 28 — Aug 15, 1986

Lawrence Berkeley Laboratory
Berkeley, California

Edited by
IAN HINCHLIFFE

Organizing Committee: **R. Cahn**
M. Davis
M. K. Gaillard
I. Hinchliffe
J. Silk

World Scientific

Published by

World Scientific Publishing Co Pte Ltd.
P. O. Box 128, Farrer Road, Singapore 9128

Library of Congress Cataloging-in-Publication data is available.

COSMOLOGY AND PARTICLE PHYSICS

ISBN 9971-50-213-5
9971-50-215-1 pbk

Printed in Singapore by Kim Hup Lee Printing Co. Pte. Ltd.

Preface

This book contains the Proceedings of the Workshop on Quarks and Galaxies held at Lawrence Berkeley Laboratory from July 28 to August 15, 1986. The workshop attendees consisted of approximately equal numbers of theoretical high energy physicists and astrophysicists. The informal atmosphere of the workshop was intended to promote discussion of topics of common interest. The workshop was supported by the Lawrence Berkeley Laboratory, the National Science Foundation under grant number PHY-8604695 and the National Aeronautics and Space Administration under grant number NGR 05-003-578. We are grateful to these agencies for providing the support which made the workshop possible. The organizing committee consisted of Professors Marc Davis and Joe Silk of the Astronomy Department at the University of Califorina at Berkeley and Prof. Mary K. Gaillard and Drs. Robert Cahn and Ian Hinchliffe of the Lawrence Berkeley Laboratory.

As will be apparent from the table of contents, the topic which excited most interest was that of the dark matter in the universe; what does it consist of and can it be responsible for the density fluctuations which lead to galaxy formation? There was also considerable interest in the problem of the lack of neutrinos being produced by the sun. The Dark Matter problem is of keen interest to both particle physicists and astrophysicists since there are particles in the currently fashionable particle physics models (such as photinos) which could be the dark matter and for which there is no terrestrial experimental evidence. In the case of the solar neutrino problem, most particle physicists would prefer it to be solved by some form of neutrino oscillations (see the paper by S. Parke), but it appears that the alternative model involving particles trapped by the sun adjusting the temperature of the core (see the talk by J. Faulkner), may be more appealing to the astrophysicists since it may also be able to solve some problems with the seismology of the sun.

There was also some discussion of inflation and whether the density fluctuations produced by it are adequate to initiate the formation of galaxies and clusters of galaxies. The search for a model which includes the field responsible for inflation in a natural way is continuing.

No workshop of this type can run smoothly without excellent technical support. I am therefore very grateful to Peggy Little, Betty Moura, Kathie Hardy and Dominique Gaillard for their invaluable assistance.

Ian Hinchliffe

This workshop was supported in part by the Director, Office of Energy Research, Office of High Energy and Nuclear Physics, Division of High Energy Physics of the U.S. Department of Energy under Contract DE-AC03-76SF00098 and in part by the National Aeronautics and Space Administration under grant NGR 05-003-578 and the National Science Foundation under grant number PHY-8604695.

Table of Contents

*Printed version not available.

COSMOLOGICAL ASPECTS OF SUPERSTRING MODELS

Pierre Binétruy

LAPP, BP. 909, 74019 Annecy-le-Vieux Cedex, France

ABSTRACT

I consider more specifically the cosmological aspects of supersymmetry breaking in "superstring models" (grand unified models which are believed to describe the effective theory obtained by compactification of superstring theories). The most interesting aspects are related to the presence of flat directions in the scalar potential (vacuum degeneracies). These flat directions are discussed both in the hidden sector of these models (do they give rise to inflation?) and in the observable sector of quarks, leptons and Higgs particles, in connection with baryogenesis.

Superstring theories[1] being theories of gravity, the possibility of testing them in the laboratory seems remote, as long as the effective low-energy theory is not known in its full details (the scales in consideration here are some 16 orders of magnitude below the typical superstring scale, of the order of the Planck mass M_{Pl}). It even remains possible that the low energy theory is basically the standard Glashow-Weinberg-Salam model (or rather its "supersymmetric" version), all other superstring remnants decoupling at scales much larger than 1 TeV. If this is the case, the early universe might prove to be a useful place to study the behaviour of these theories at scales intermediate between 1 TeV and the Planck mass. Probably the most spectacular effect of superstring theories on the evolution of the early universe is the compactification of the six extra dimensions.

2

Let me recall briefly[2] that superstring theories must be defined in the critical dimension of 10, if one wants to maintain conformal invariance on the world-sheet[*]. The different physical states are associated with oscillations of the string. In particular, the bosonic massless states are : the graviton, a scalar field which is a dilaton and an antisymmetric tensor field. The field theory limit is a Yang-Mills theory coupled to N=1 supergravity in 10 dimensions. The gauge group is almost uniquely determined by the requirement of the cancellation of 10-dimensional gauge and gravitational anomalies[1]. As is well-known by now, this restricts the choice to $E_8 \times E_8$ and SO(32). SO(32) has phenomenological problems of its own (it yields only vector-like representations[4]) and I will concentrate upon $E_8 \times E_8$.

Clearly, during the early evolution of the Universe, compactification of six dimensions must occur. The 10-dimensional manifold is therefore effectively the product of a 4-dimensional Minkowski space-time and a 6-dimensional compact manifold K. It is a desirable feature to conserve supersymmetry through the process of compactification. The reason (which is unfortunately only a technical one) is that supersymmetry is the only known way to account for the hierarchy problem : why is the low energy effective theory stable under radiative corrections when scales of order M_{Pl} are present? It turns out[5] that, in the case where supersymmetry is conserved down to low energy, large radiative corrections from superheavy particles cancel between supersymmetric partners. The requirement that one supersymmetry remains intact through compactification imposes in turn constraints on the choice of the compact manifold. Two favourite candidates seem to have emerged in the last year : Calabi-Yau manifolds[4] and orbifolds[4,6]. After compactification, one ends up[4] with a 4-dimensional Yang-Mills theory, the gauge group being of the form K × G subgroup of $E_8 \times E_8$. Matter fields are singlets under K ; the only fields non-singlet under K are the gauge fields of K (in the adjoint representation) and their supersymmetric partners (gauginos). They interact with the matter fields only through gravitational interactions and therefore form

---

[*] An evolving string spans a 2-dimensional surface, known as the world sheet. Conformal invariance[3] is needed in order that the longitudinal modes of the string decouple from physical amplitudes.

a hidden sector. Ordinary quarks and leptons are to be found among the matter fields, non-singlet under G. Therefore, G must include the known gauge interactions $SU(3) \times SU(2) \times U(1)$. In the case of a compactification on a Calabi-Yau manifold, G is E_6 or one of its subgroups. It is useful in this case to describe the field content in terms of the underlying E_6 symmetry. The matter fields are in N_G (number of generations) complete 27 representations plus pieces of $27 + \overline{27}$ (determined once is known the way E_6, or rather E_8, is broken to G).

In the description above, I have distinguished three levels and correspondingly three mass scales : i) the string level at a scale $M_s = T^{1/2}$ (T string tension) ii) the (10-dimensional) field theory limit characterized by the Planck scale M_{P1} iii) its compactified (4-dimensional) version at a scale $M_{comp.}$; $M_{comp.}$ is also the scale where all the gauge couplings are equal : we therefore have effectively a grand unification of the gauge interactions and $M_{comp.} = M_{GUT}$. It is not accurate however to separate three stages because all the scales turn out to be of the same order:

$$M_s \sim M_{P1} \sim M_{comp.} \sim M_{GUT} \tag{1}$$

The reason[7,8] is that in a string model there is only one dimensionless parameter. This parameter can be chosen to be the gauge coupling constant at grand unification, whose magnitude is fixed by phenomenology. This in turn imposes the constraint (1). To be complete, one must say that compactification introduces other dimensionless parameters and one has to make some other reasonable assumptions[7] (e.g. the string theory is not strongly coupled) to reach the same conclusion.

Unfortunately, a constraint such as (1) relates the process of compactification to the dynamics of string ($M_{comp.} \sim M_s$) and as long as this dynamics is not known at the quantum level (string field theory?) it is difficult to study the evolution of the universe during this epoch. Similar difficulties arise from $M_{comp.} \sim M_{P1}$ because we do not know yet the exact form of the gravity theory that emerges from superstrings (higher order terms in the Lagrangian R^2, $R_{MN} R^{MN}$ ...).

In the following, I will concentrate upon a phenomenon of basic importance for low energy phenomenology and which occurs at scales smaller than the common scale (1) : the breaking of supersymmetry. As mentioned earlier, one supersymmetry must remain intact after compactification if one wants to avoid the hierarchy problem. The effective theory below $M_{comp.}$ is therefore a 4-dimensional Yang-Mills theory with matter fields coupled to N=1 supergravity. At which scale is supersymmetry broken and in which way? This question has been much studied in the general case of supergravity but the models that emerge from superstrings ("superstring models") have some special features that make it a rather delicate matter to study.

We know that supersymmetry is broken at low energy and finding the supersymmetric partners of quarks and leptons would give some first hand information. But cosmology can also give us some valuable clues about supersymmetry-breaking. And in that respect flat directions of the scalar potential (vacuum degeneracies) play a special role : they are lines (surfaces) in field space, running to infinity, along which the potential V is zero.

Certainly, in a non-supersymmetric theory, no one would pay attention to these flat directions : radiative corrections lift the corresponding vacuum degeneracy and the so-called "flatness" is only a tree-level property which has no intrinsic meaning. This is not the case in a supersymmetric theory and one can prove[9], in connection with the non-renormalisation theorem[5], that flat directions remain flat, to all orders of perturbation. It follows that the appearance of a structure in a flat direction is a sign of supersymmetry-breaking. In supergravity theories, the order parameter for supersymmetry-breaking is a priori the gravitino mass $m_{3/2}$. Therefore the details of the structure are in principle of order $m_{3/2}$. One should be careful however in the case where there is a hidden sector. Strictly speaking, $m_{3/2}$ determines the scale of supersymmetry-breaking in the hidden sector. It is possible (and it is indeed the case in superstring models) that a different scale $\tilde{m}$ sets the scale of supersymmetry-breaking in the observable sector of quarks and leptons (G sector, in our notations). Of course, $\tilde{m}$ is a function of $m_{3/2}$ (typically $\tilde{m} = m_{3/2}^{n}/M_{Pl}^{n-1}$) since

the signal of supersymmetry breaking is a massive gravitino. The mass
of the Higgs scalar responsible for $SU(2) \times U(1)$ breaking is also
given by $\tilde{m}$. Since we know at which scale this breaking occurs, we can
put a severe constraint on $\tilde{m}$:

$$\tilde{m} = O(1 \text{ TeV}) \tag{2}$$

If a flat direction appears in the potential of observable scalars
(Higgs), the structure that is induced in this direction by supersymmetry breaking scales as $\tilde{m}$. If this determines a ground state for
which the vacuum expectation values (vevs) are much larger than 1 TeV
(as in the case in superstring models), we obtain almost flat directions in the scalar potential. It is well-known that such a configuration can have striking effects on the evolution of the Universe :
for example, the evolution of a scalar field along such a direction
might lead to an inflationary epoch or conversely (depending on the
time at which it occurs) to a late release of entropy that endangers
the successes of the standard Big Bang scenario.

I will in the following concentrate upon the cosmological evolution associated with these flat directions. It is at present one of
the best ways to study supersymmetry-breaking in superstring models.
As I will discuss in Section 1, it is of particular relevance in these
models because the question of supersymmetry-breaking is still an open
one (in particular the way it is transferred from the hidden to the
observable sector). I then discuss in Section 2 flat directions in the
hidden sector and whether their presence in the model leads to inflation. Finally, in Section 3, I turn to flat directions in the observable sector and study their connection with baryon number generation.

1.   SUPERSYMMETRY BREAKING IN SUPERSTRING MODELS

I will first discuss two flat directions of the scalar potential
which are of particular relevance for our discussion. From the preceding comments, it should come as no surprise that we chose to start
this way to discuss the breaking of supersymmetry.

One of these flat directions is present even at the string level.

It is associated with the dilaton field ϕ; which, as we said, is one of the massless bosonic modes of a string. It turns out[10,8] that, at the classical level, all S matrix elements have the same dependence in $\phi^{*)}$. Therefore all ϕ vacua are related to one another by a symmetry (a rescaling of ϕ amounts to a rescaling of all tree amplitudes or, in field theory language, of the Lagrangian itself). And the potential of the dilaton is flat.

The second direction is connected with compactification. There is a field T (called "breathing mode") associated with the fluctuations of the overall size of the compact manifold K. If we consider the 10-dimensional metric g_{mn} (m,n = 0,...,9), its dependence in T reads:

$$g_{mn} = \begin{pmatrix} T^{-3} g_{\mu\nu} & \\ & T g_{MN} \end{pmatrix} \tag{3}$$

with $\mu,\nu = 0,...,3$, $M,N = 4,...,9$ [**)].

The potential of T is also flat. In other words, two compact manifolds of the same shape but different size are equally preferred. The presence of such flat directions is certainly a drawback of the model. To give a hint of the problems that arise, let me write some terms of the Lagrangian in four dimensions:

$$\frac{1}{e} \mathcal{L}^{(4)} = S \left(-\frac{1}{2} R^{(4)} - \frac{1}{4} F_{\mu\nu}^2 + ... \right) \tag{4}$$

where

$$S = T^3 \phi^{-2}, \tag{5}$$

$R^{(4)}$ is the curvature of the 4-dimensional space-time and $F_{\mu\nu}$ the Yang-Mills field strength.

If S has a vanishing potential, the first term in $\mathcal{L}^{(4)}$ tells us

---

*)For the simplicity of the discussion, I am working until further notice with units where the string scale is 1 (see ref.8 for example).

**)The dependence of the 0...3 components of g_{mn} on T is only a consequence of the fact that we chose to have the same value for the Planck mass in 4 and 10 dimensions.

that S is a Brans-Dicke field with ω of order 1. This is ruled out. Also from the second term in (4), we find that the gauge coupling (in 4 dimensions) is given by

$$\frac{1}{g^2} = S. \tag{6}$$

If S has no definite ground state, g has no preferred value and the model is not determined.

We said earlier that such a degeneracy is not lifted, to all orders of perturbation theory. We therefore need to include some non-perturbative effects.

Affleck, Dine and Seiberg[11] have made a thorough analysis of supersymmetry breaking through non-perturbative effects. The review of numerous cases led them to empirically conclude that only two cases are possible :

i) the degeneracy is not lifted by non-perturbative effects.

ii) a non-zero potential is generated which slopes to zero at infinity (from above or from below).

For example, in the case of SU(N) with matter fields in the adjoint representation, flat directions are not lifted (i). On the other hand, an example of case ii) is a gauge theory with a singlet field S coupling to the gauge field[*]:

$$\frac{1}{e} \mathcal{L} = -\frac{1}{4} S F^{\mu\nu} F_{\mu\nu} - \frac{1}{2} \bar{\lambda} \not{D} \lambda \tag{7}$$

where the λ are the gauginos, supersymmetric partners of the gauge fields. This is precisely the case of superstring models (compare (7) with (4)). It is easy to obtain the form of the potential for S. At a scale

$$\Lambda_c \sim e^{-1/(2b_o g^2)} \sim e^{-S/(2b_o)} \tag{8}$$

the gauge interactions in the hidden sector become strong (b_o is the coefficient of the beta-function for the group K ; I also take S —

---

[*] From now on, we turn to units where $M_{Pl} = 1$.

and T - to be real fields : their imaginary parts have very interesting properties but I will not discuss them here). Therefore the gauginos should condense at scale Λ_c :

$$<\bar\lambda\lambda> \sim \Lambda_c^3 = e^{-3S/2b_0} \tag{9}$$

and the four-gaugino coupling present in the theory[12] yields a non-
trivial potential for S.

$$V_{eff} \sim |<\bar\lambda\lambda>|^2 \sim e^{-3S/2b_0} \tag{10}$$

What about supersymmetry-breaking in this context? Let me recall that,
in supergravity theories, supersymmetry is broken spontaneously through
the superHiggs mechanism[13]. In the well-known Higgs mechanism, a
gauge boson eats up a Goldstone boson to become massive (degrees of
freedom 2+1 = 3). Goldstone bosons are scalar fields which can be
gauged away. Similarly, in the superHiggs mechanism, a massless gravitino eats up a Goldstone fermion - Goldstino - to become massive
(degrees of freedom : 2+2 = 4 in 4-dimensional space-time). To find
a Goldstino, one has to look for a fermion field χ which can be trans-
formed away by a supersymmetry transformation ($\delta\chi$ = cst × ε + $\cdots$
where ε is a constant spinor associated with the transformation). For
instance, the supersymmetric partner of a scalar field transforms as[12] :

$$\delta\chi = \bar\lambda\lambda \, \varepsilon + \cdots \tag{11}$$

Therefore gaugino condensation ($<\bar\lambda\lambda>$ = cst) induces the presence of a
Golstino and hence supersymmetry breaking[14]. The gravitino mass reads
$m_{3/2} = \Lambda_c^3/M_{Pl}^2$ but unfortunately a large cosmological constant $O(\Lambda_c^4)$
is also generated. It is of primary importance however that our way of
breaking supersymmetry does not induce a cosmological constant. The
vacuum energy seems to be zero at the level of strings[15] and, if supersymmetry is to be broken after compactification (i.e. in 4 dimensions),
it must leave intact this remarkable property.

Luckily enough, there is another way to break supersymmetry. I
mentioned earlier that the massless sector of the closed string includes

an antisymmetric tensor field. Its field strength H is involved in
the transformation of the gauginos[16)]

$$\delta\lambda = \frac{1}{8} \phi^{-3/4} \Gamma^{mnp} H_{mnp} \varepsilon + \cdots \tag{12}$$

where Γ^{mnp} are antisymmetric products of gamma matrices and notations
similar to eq.(3) are used for indices. If H_{MNP} acquires a vev
($<H_{ijk}> = c \; \varepsilon_{ijk} \neq 0$ where i,j,k refer to the three complex coordinates of the compact manifold) then we have found a Goldstino and supersymmetry is broken[17)]: a non-zero mass for the gravitino but also unfortunately a non-zero cosmological constant are generated.

Dine, Rohm, Seiberg and Witten[18)] noted however that combining the
two mechanisms allows to cancel the two contributions to the cosmological constant. Indeed, the tree level potential for S and T now
reads

$$V(S,T) = \frac{1}{16ST^3} \left| c + h(1 + 3\frac{S}{b_0}) \; e^{-3S/2b_0} \right|^2 \tag{13}$$

where the first term is the contribution of $<H>$(c), and the second
one originates from gaugino condensation (compare its exponential dependence in S with the crude estimate eq.(10)). The ground state
(S_0, T_0) is trivially obtained by writing:

$$c + h (1 + 3 \frac{S_0}{b_0}) \; e^{-3S_0/2b_0} = 0 \; , \tag{14}$$

and the gravitino mass is computed to be

$$m_{3/2} = \frac{1}{4S_0^{1/2}T_0^{3/2}} \left| c + h \; e^{-3S_0/2b_0} \right| . \tag{15}$$

Clearly, S_0 is fixed by eq.(14) but not T_0. In other words,
$V(S_0,T) = 0$ and we are left with a flat direction in T. This means
in particular that the gravitino mass remains undetermined at tree
level (it depends on T_0). We will come back in the next Section to
the determination of T_0 (and $m_{3/2}$) but we will start with the cosmological evolution associated with S.

2. FLAT DIRECTIONS IN THE HIDDEN SECTOR. INFLATION.

We are now able to describe the cosmological evolution associated with the field S.

When the Universe emerges from the compactification epoch ($T \sim M_{comp}$), gauginos have not condensed and the vev of H is zero : both of these conditions are imposed by supersymmetry; moreover if $<\bar{\lambda}\lambda> = 0$, it follows that $c = 0$ because it is the value that minimizes the total potential energy.

At a scale Λ_c, the gauginos of the hidden sector condense. This induces a non-trivial structure in the S direction of the potential (typically eq.(10)) but it also triggers a non-zero vev for H : this time because, since $<\bar{\lambda}\lambda> \neq 0$, the total energy is minimized by a non-zero c. Rohm and Witten[19] have studied in detail this process, which requires some attention. The point is that, because H is a field strength (in the language of differential forms, H = dB, where B is the antisymmetric tensor of the massless bosonic sector of superstrings), c can only take discrete values. Indeed, it can be written as an integral over a closed oriented 3-surface S :

$$c = \int_S H^{MNP} \, d\Sigma_{MNP} \tag{16}$$

which is quantized in much the same way as the charge of a monopole is quantized. If this was the end of the story, it would be a disaster because c would be fixed once for all (by the topology of the compact manifold) and would not be allowed to change from zero to a non-zero value. It turns out[16,4] that there are in H more terms than simply the differential of B, dB : they are known as the Chern-Simons terms. These terms allow tunneling from one quantized value for c to another one[19]. It is therefore possible for the Universe to evolve to a phase where the value of c minimizes its total energy. Because the process is not known in detail, we have to suppose that it occurs fairly quickly. If this was not the case, some energy of order Λ_c^4 would be stored in the vacuum for a non negligeable amount of time, which would lead to inflation. We would be left with the problems of the original scenario[20] for inflation (since there is tunneling).

Once the Universe has evolved to the $c \neq 0$ phase (according to our hypotheses, this should occur at $T \lesssim \Lambda_c$), the potential for S and T is given by eq.(13). It is then an interesting question to determine whether the structure in the S direction satisfies the conditions for a new inflation scenario[21]. Before addressing it, let me give some arguments why inflation should occur after compactification.

Let me define μ_o as the typical scale of the inflaton potential (the inflaton is the field whose evolution in the potential leads to inflation) : the energy stored in the vacuum is of order μ_o^4.

The production of gravitons during the de Sitter phase leads to relic gravitational waves at present time and therefore to distortions in the $3°K$ radiation background. This allows to put restrictions on the scale μ_o[22]:

$$(\frac{\Delta T}{T})^2 \sim (\frac{\mu_o}{M_{Pl}})^4 < 10^{-8} \tag{17}$$

Analogous constraints[23] are obtained from the requirement that density fluctuations in the de Sitter phase should lead to galaxy formation without giving too large anisotropies of the radiation background. We see that μ_o has to be at least two or three orders of magnitude smaller than the common scale $M_{Pl} \sim M_{comp} \sim M_{GUT}$ (eq.(1)).

The same conclusion can be reached using other arguments. For example, Wen and Witten[24] have shown that in the process of compactification from 10 to 4 dimensions, monopoles are created (as we said earlier, part of the gauge symmetry is broken at the same time). Typically, the ratio of the number of such monopoles to entropy is given by

$$\frac{n_{mon}}{S} > (\frac{T_c}{M_{Pl}})^3 \sim O(1) \tag{18}$$

since the critical temperature is given here by M_{comp} ($\sim M_{Pl}$ from (1)). The only known way to dilute these monopoles is to have an inflationary epoch following compactification ($\mu_o < M_{comp}$).

The natural question is therefore to ask whether the S field

12

could be responsible for such a behaviour. If this was the case, the scale μ_o would be of order $m_{3/2}^{1/2}$ (compare eqs (13) and (15)) and for example $\mu_o/M_{Pl} \sim 10^{-3}$ for $m_{3/2} \sim 10^{13}$ GeV. The study of the potential (13) in connection with inflation has been done in Ref. 25) (and, under somewhat different hypotheses but with similar conclusions, in Ref. 26)). The shape of the potential at $T = 0$ and $T = M_{Pl}$ *) depends on the value of $\hat{c} = c/h$ and is given in Fig. 1 in terms of $\alpha = 3S/b_o$ (see Ref. 25). The ground state at high temperature determines[27) the initial condition. Clearly in order to have inflation, the field must lie away from its minimum (minima) at zero temperature. This occurs only in cases (b) and (c) of Fig. 1, that is $0.937 \leq \hat{c} < 1$. In case b) inflation can proceed in the standard way[21) whereas case c) involves tuneling and requires a scenario à la Hawking and Moss[28), which further restricts the constraint on $\hat{c}$. We find[25)

$$\hat{c} = 0.9374 + \delta\hat{c} , \quad \delta\hat{c} < 3 \times 10^{-4} . \tag{19}$$

This is definitely much too strong a constraint. We have noted earlier that $\hat{c}$ is uniquely determined by the topology of the compact manifold, and as long as we have not exhibited one which yields precisely the value (19), we must conclude that inflation does not seem to occur in the direction of S.

What about the T field? We left it at the end of Section 1 with a flat potential. It has been realized for some time that a nontrivial potential is generated by radiative corrections at the one-loop level[29,30) but which structure arises is still a matter of discussion and I will only summarize the situation here.

In some cases[30), the potential obtained is unbounded from below towards T = 0. This situation is intolerable and some non-perturbative effects (which, we know, do play a rôle in the $T \sim 0$ regime) must stabilize the potential. However, the potential for $T \to \infty$, where we trust the perturbation expansion, is negative; this seems to indicate

---

*)Our methods for computing the effective potential at high temperature are certainly not valid at $T = M_{Pl}$ but this gives a trend and the potential at realistic temperatures is obtained by the appropriate rescaling.

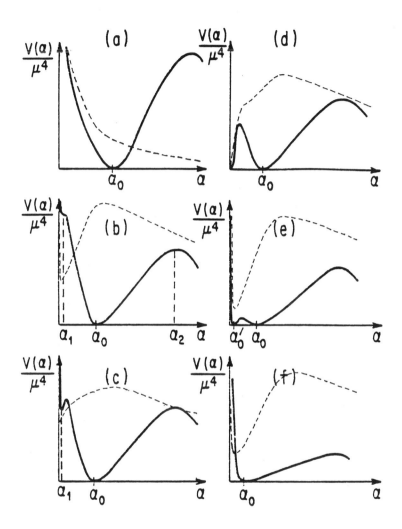

Fig. 1 : Shape of the potential $V(S)$ (eq.(13); $\alpha = 3S/b_0$)
at temperature $T = 0$ (solid line) and $T = M_{Pl}$
(dashed line ; the scale is different) for different
values of $\hat{c} = c/h$: (a) $\hat{c} < 0.937$ (b) $\hat{c} = 0.937$
(c) $0.937 < \hat{c} < 1$ (d) $\hat{c} = 1$ (e) $1 < \hat{c} < 1.21$
(f) $\hat{c} = 1.21$.

that any ground state that we would obtain this way would yield a non-zero (negative) cosmological constant. Maeda, Pollock and Vayonakis[31] have devised a clever scenario for inflation based on such a stabilized potential (actually the one originally given in Ref. 29). But they have to add a constant term to cancel the negative cosmological constant, which seems to be a rather ad hoc treatment of a fundamental problem. In particular, we have stressed earlier that, if supersymmetry-breaking is a post-compactification process, it seems difficult to invoke higher string modes to cancel the cosmological constant thus created.

In a recent work[32], we have adopted a somewhat different attitude. A natural cut-off for the (supersymmetry-breaking) radiative corrections is provided by the condensation scale Λ_c (eq.(9)), above which the gaugino condensates are broken into their constituents. This is not quite the case when we deal with a vev like $<H>$ and we argue in Ref. 32 that the corresponding contributions should be cut-off at $M_{comp} \sim M_{Pl}$. Under this hypothesis and for a number of chiral fields slightly larger than the number of gaugino fields, we obtain a stable potential which slopes down (from above) to zero at $T \to \infty$. The potential energy is in this case positive (a good point for inflation!). We further argue that, whenever there is a ground state for finite T, the corresponding cosmological constant is zero (at least at the one-loop level that we are considering here). The potential thus obtained is therefore a perfectly valid tool to discuss inflation. A detailed analysis remains to be performed.

3.   FLAT DIRECTIONS IN THE OBSERVABLE SECTOR. BARYOGENESIS.

We now turn to the observable sector of superstring models, that is those fields which are in 27 and $\overline{27}$ of E_6.

As we discussed earlier, the supersymmetry-breaking parameter $\tilde{m}$ is fixed by phenomenology to be of order 1 TeV and need not be equal to the gravitino mass. This is indeed the case in superstring models. We discussed at the end of Section 2 how the field T is determined by radiative corrections, which in turn fixes the gravitino mass (eq.(15)). In the observable sector, even after supersymmetry-breaking, scalars and gauginos remain massless at tree level : this illustrates the difficulty of sending the information of supersymmetry breakdown from the hidden to the observable sector (a general feature to all no-scale models[33]). At the one-loop level however, although scalars remain massless[29], gauginos acquire a non-zero mass[34] and supersymmetry is therefore effectively broken in the observable sector. This in turn yields a non-zero mass for scalars, which we will define as our effective supersymmetry breaking parameter $\tilde{m}$.

As noted earlier, almost flat structures will appear in the potential, in the direction of the fields whose vev is much larger than $\tilde{m}$. This restricts our analysis to $SU(3) \times SU(2) \times U(1)$ gauge singlets.

There are two $SU(3) \times SU(2) \times U(1)$ singlets in a 27 of E_6 : one of them is a singlet under $SO(10)$ (we will denote it by N) and among fields charged under $SO(10)$, one (N') is a singlet under $SU(5)$. We can safely consider N and N' on their own since they are the only fields which can acquire a large vev. There are $N_G + \delta$ of them, δ being the number of fields with opposite quantum numbers $(\bar{N}, \bar{N}')$ coming from the $\overline{27}$. We will consider for simplicity only one fo them (say N).

It is easy to check, from E_6 symmetry requirements, that no renormalizable term can be constructed in the superpotential W that involves only N or N'. The corresponding F-terms in the potential $(|dW/dN|^2)$ are zero. Since the D-terms are of the type $(N^2 - \bar{N}^2)^2$, the potential is zero for $N = \bar{N}$. This is the flat direction that we are looking for.

The corresponding vacuum degeneracy is lifted by:

i) non-renormalizable terms in the superpotential $(N\bar{N})^n/M_{P1}^{2n-3}$ $(n \geq 2)$

ii) supersymmetry breaking (for simplicity, we only include here a mass term).

The potential reads typically in the flat direction (zero D-terms):

$$V = \tilde{m}^2 N^2 + \lambda N^{4n-2}/M_{P1}^{4n-6} \ . \tag{20}$$

The ground state is obtained for:

16

$$N_o/M_{P1} \sim (\tilde{m}/M_{P1})^{1/(2n-2)}$$

(21)

$$\mu_o^4/M_{P1}^4 \equiv (V(0)-V(N_o))/M_{P1}^4 \sim (\tilde{m}/M_{P1})^{(2n-1)/(n-1)}.$$

This yields typically $N_o \sim 10^{10}$ GeV, $\mu_o \sim 10^6$ GeV for $n = 2$ but symmetries[35] can prevent $n = 2$ terms to be present in the super-potential, thus pushing N_o to larger scales.

The evolution with temperature of the potential for N has first been discussed by Yamamoto[36] and is summarized in Fig. 2.

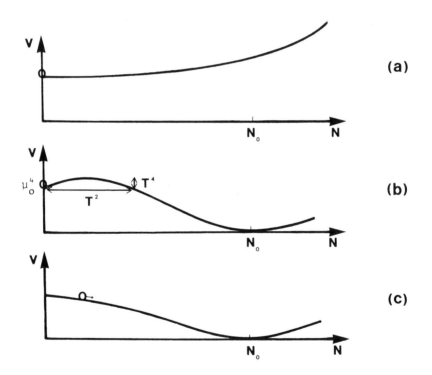

Fig. 2 : Evolution of the potential for N with temperature :
(a) $T \gg N_o$ (b) $\tilde{m} < T < N_o$ (c) $T < \tilde{m}$.

At very high temperature, the gauge symmetry associated with N is restaured (N is not an E_6 singlet) and the ground state is $<N> = 0$. As temperature decreases, a dip appears at $N = N_o$ and N_o soon

becomes the absolute minimum : temperature corrections are exponentially suppressed at large N (this is always the case but it is here of
special importance because $N_o \gg \mu_o$). For $\tilde{m} < T < N_o$, a barrier of
height T^4 separates the origin from the true minimum N_o and the
field is trapped at the origin[*]. For $T < \tilde{m}$, this barrier disappears
and the field rolls down (fairly quickly) to N_o.

Since $\mu_o > \tilde{m}$, it is clear that in the temperature range
$\tilde{m} < T < \mu_o$, the energy stored in the vacuum dominates the total energy
density of the Universe and triggers some exponential expansion. This
inflationary epoch is however not long enough to cure all the problems
that inflation is supposed to solve. Indeed it yields[25] a maximum of
20 e-foldings in the cosmic scale factor increase, whereas at least 65
are needed[20]. Once the field has started falling, it oscillates
around its ground state N_o. Because it is weakly coupled to matter[**],
there will be a large release of entropy (a factor of entropy increase
larger than 10^{12}, see Ref. 36) at a late time in the evolution of
the Universe. All existing concentrations are considerably diluted
away, in particular the baryon density. Such an evolution requires
therefore some mechanism for generating baryon asymmetry that works at
a low temperature (after the decay of the N field, the Universe re-
heats to a temperature $T_{RH} \sim (\Gamma M_{Pl})^{1/2}$).

The standard ingredients for generating baryon number are[38] :
baryon number violating interactions, a CP violating phase and a distribution out of equilibrium. The first requirement points towards one
specific component of the 27 : the g-quarks which are color triplets,
$SU(2)_L$ singlets (their scalar partners can be interpreted as the color
triplet Higgs of the grand unified theories). Their $(E_6$-allowed) interactions can generate proton decay so that, if these g quarks are
light, one has to forbid such couplings : apart from the unnaturalness
of such a choice (no explicit model has been constructed which incorpo-
rates these constraints), it is difficult to see how one could create a

---

[*] Except if something – such as inflation – happens in the mean time
that flattens out temporarily the barrier[37].

[**] The coupling is of order $1/N_o$ since it is mediated by heavy particles of mass N_o times a coupling constant. Hence the corresponding
decay rate is $\Gamma \sim \tilde{m}^3/N_o^2 \ll 1$.

non-zero baryon number in the Universe.

Fortunately, the coupling gg^cN which is present in the superpotential allows to give a large mass to the g quarks. If this mass
is large enough, we can keep the baryon-violating interactions of the
g-quarks without having a fast proton decay.

To obtain an out of equilibrium distribution, Yamamoto[36] proposes
to use a mechanism which was originally devised for inflationary scenarios[39]. The N field decays into particles of mass much larger than
T_{RH} : they are thus evidently out of equilibrium.

A clever mechanism has also been proposed by Lazarides,
Panagiotakopoulos and Shafi[40]. They note that when $<N> = 0$, the g-
quarks have a mass $\tilde{m}$, whereas for $<N> \neq 0$, they acquire a much
larger mass of order $<N>$ (through the coupling gg^cN). We said that
at $T = \tilde{m}$, the field N falls rapidly from the origin to its ground
state N_o ; accordingly, the mass of the g quark blows up from $\tilde{m}$
to N_o, too quickly for its thermal distribution to follow : the g-
quarks are out of equilibrium.

Finally, the mechanism proposed by Affleck and Dine[41] is another
striking example of the richness of scenarios using flat directions of
the potential. Consider such a direction ϕ not connected with gauge
symmetry breaking. They argue that the field ϕ can be left by an
inflationary expansion quite far off the origin (by fluctuations in the
de Sitter phase). When supersymmetry is broken, some structure develops
along this direction and the field ϕ falls towards its ground state.
Through baryon number and CP violating quartic couplings, this can generate a significant baryon density. This scenario has been worked out
in detail in the case of superstring models by B. Campbell, J. Ellis,
D.V. Nanopoulos and K. Olive[42].

To conclude briefly, the cosmology of flat directions of the potential is a unique way of having access to the details of supersymmetry
breaking. This is of special importance for superstring models where
this breaking might occur at a scale intermediate between low energy
(1 TeV) and the Planck scale. At this point, it is not so much definite
testable predictions than the possibility of writing a self-consistent

and complete cosmological scenario that will put strong constraints on superstring models. The issues at stake are for example the transition from $\langle H \rangle = 0$ to $\langle H \rangle \neq 0$ (and is $\langle H \rangle$ equal to zero at the end of compactification?), the possibility of implementing inflation, the question of stability of the proton vs. baryogenesis...

ACKNOWLEDGEMENTS

I wish to thank the organizers of the Workshop "Quarks and Galaxies" for providing us with a very relaxed and stimulating atmosphere. And I am extremely grateful to Mary K. Gaillard and all the members of the Theory Group at LBL for the hospitality extended to me this summer and the numerous lively and very fruitful discussions that I enjoyed with them.

REFERENCES

1. Green M. and Schwarz J., Nucl. Phys. B218, 43 (1983); Phys. Lett. B149, 117 (1984);
    Gross D., Harvey J., Martinec E. and Rohm R., Phys. Rev. Lett. 54, 502 (1985); Nucl. Phys. B256, 253 (1985); Nucl. Phys. B267, 75 (1986).

2. Schwarz J.H., Phys. Rep. 89, 223 (1982);
    Green M., Surveys in High Energy Physics 3, 127 (1984).

3. See for example D. Friedan, in Proceedings of the 1982 Les Houches Summer School, J.B. Zuber and R. Stora eds (North Holland 1984).

4. Candelas P., Horowitz G., Strominger A. and Witten E., Nucl. Phys. B258, 46 (1985).

5. Grisaru M., Siegel W. and Roček M., Nucl. Phys. B159, 429 (1979).

6. Dixon L., Harvey J., Vafa C. and Witten E., Nucl. Phys. B261, 678 (1985); Nucl. Phys. B274, 285 (1986).

7. Dine M. and Seiberg N., Phys. Rev. Lett. 55, 366 (1985); Phys. Lett. 162B, 299 (1985);
    Kaplunovsky V., Phys. Rev. Lett. 55, 1036 (1985).

8. Seiberg N., Lectures at the 3rd Jerusalem Winter School - Strings and Superstrings (1985).

9. Witten E., Nucl. Phys. B188, 513 (1981).

10. Witten E., Phys. Lett. 155B, 151 (1985).

20

11. Affleck I., Dine M. and Seiberg N., Nucl. Phys. **B256**, 557 (1985).

12. Cremmer E., Ferrara S., Girardello L. and Van Proeyen A., Nucl. Phys. **B212**, 413 (1983).

13. Deser S. and Zumino B., Phys. Rev. Lett. **38**, 1433 (1977).

14. Nilles H.P., Phys. Lett. **115B**, 193 (1982); Ferrara S., Girardello L. and Nilles H.P., Phys. Lett. **125B**, 457 (1983).

15. Moore G., Harris J., Nelson P. and Singer I., Phys. Lett. **178B**, 167 (1986); Chang M.C. and Ran Z., "Divisors on $\bar{M}_g$ and the Cosmological Constant", Univ. of So.Carolina/U.C. Riverside Math. Dept. Preprint (1986).

16. Chapline G. and Manton N., Phys. Lett. **120B**, 105 (1983).

17. Derendinger J.P., Ibáñez L. and Nilles H.P., Phys. Lett. **155B**, 65 (1985).

18. Dine M., Rohm R., Seiberg N. and Witten E., Phys. Lett. **156B**, 55 (1985).

19. Rohm R. and Witten E., Ann. of Phys. **170**, 454 (1986).

20. Guth A., Phys. Rev. **D23**, 347 (1981).

21. Linde A.D., Phys. Lett. **108B**, 389 (1982); Albrecht A. and Steinhardt P.J., Phys. Lett. **140B**, 44 (1984).

22. Rubakov V.A., Sazhin M.V. and Veryaskin A.V., Phys. Lett. **115B**, 189 (1982).

23. Hawking S.W., Phys. Lett. **115B**, 295 (1982); Starobinsky A.A., Phys. Lett. **117B**, 175 (1982); Guth A. and Pi S.Y., Phys. Rev. Lett. **49**, 1110 (1982); Bardeen J.M., Steinhardt P.J. and Turner M.S., Phys. Rev. **D28**, 679 (1983).

24. Wen X.G. and Witten E., Nucl. Phys. **B261**, 651 (1985).

25. Binétruy P. and Gaillard M.K., Enrico Fermi Institute, Berkeley preprint EFT 86-30, LBL-21621, UCB-PTH-86/15 (1986) (to be published in Phys. Rev. D).

26. Ellis J., Enqvist K., Nanopoulos D.V. and Quiros M., CERN preprint TH.4325 (1985).

27. For a discussion of initial conditions, see for example Binétruy P., Supergravity and Inflation, in Proceedings of the 6th Workshop on Grand Unification, Minneapolis, April 18-20, 1985 (World Scientific 1986) p.403.

28. Hawking S. and Moss I., Phys. Lett. 110B, 35 (1982).

29. Breit J.D., Ovrut B.A. and Segré G., Phys. Lett. 162B, 303 (1985); Binétruy P. and Gaillard M.K., Phys. Lett. 168B, 347 (1986).

30. Ahn Y.J. and Breit J.D., Nucl. Phys. B273, 75 (1986); Quiros M., Phys. Lett. 173B, 265 (1986).

31. Maeda K., Pollock M.D. and Vayonakis C.E., Trieste ICTP Preprint IC/86/5 (1986).

32. Binétruy P., Dawson S., Hinchliffe I. and Gaillard M.K., Berkeley preprint LBL-22339, UCB-PTH-86/31 (October 1986) and paper in preparation.

33. Cremmer E., Ferrara S., Kounnas C. and Nanopoulos D.V., Phys. Lett. 133B, 61 (1983); Ellis J., Kounnas C. and Nanopoulos D.V., Nucl. Phys. B241, 406 (1984); Nucl. Phys. B247, 373 (1984); Phys.Lett. 143B, 410 (1984).

34. Binétruy P., Dawson S. and Hinchliffe I., Phys. Lett. 179B, 262 (1986) and Berkeley preprint LBL-22322, UCB-PTH-86/34 (Oct. 1986).

35. Greene B., Kirklin K., Miron P. and Ross G., Phys. Lett. 180B, 69 (1986).

36. Yamamoto K., Phys. Lett. 168B, 341 (1986).

37. Bertolami O. and Ross G.G., private communication.

38. See for example : Kolb E. and Turner M., Ann. Rev. Nucl. Part. Sci. 33, 645 (1983).

39. Dolgov A.D. and Linde A.D., Phys. Lett. 116B, 329 (1982); Abbott L., Farhi E. and Wise M.B., Phys. Lett. 117B, 29 (1982).

40. Lazarides G., Panagiotakopoulos C. and Shafi Q., Phys. Rev. Lett. 56, 557 (1986).

41. Affleck I. and Dine M., Nucl. Phys. B249, 361 (1983).

42. Campbell B., Ellis J., Nanopoulos D.V. and Olive K., CERN preprint TH.4484/86 (July 1986).

STRUCTURE FORMATION IN $\Omega = 1$ CDM UNIVERSES
AND THE BIASING MECHANISM

J. Richard Bond

Canadian Institute for Theoretical Astrophysics
Toronto, CANADA

ABSTRACT

The properties of structures forming in $\Omega = 1$ $h = 0.5$ universes dominated by cold dark matter (CDM) are considered using the theory of peaks of primordial density fields, assuming they are initially Gaussian. Four such models are considered: In the conventional adiabatic model with Zeldovich initial conditions, for Ω_X to be approximately one galaxies must be more clustered than the mass distribution. A tentative physical explanation for why bright galaxies would be likely to form only at high peaks of the primordial density field is presented. Peaks collapse first to protogalactic pancakes, flattened clusters of dwarf galaxies which lose gas while expansion in the pancake plane continues. Low amplitude peaks tend to expand for much longer times than high ones. This biasing mechanism is shown quantitatively to be sufficient to reconcile $\Omega = 1$ with dynamical estimates of the cosmological density parameter. This model disagrees with observational determinations of large scale streaming velocities and the cluster-cluster correlation function. A simple ad hoc modification in the scale-invariant initial conditions with extra power on large scales is shown to preserve the adiabatic CDM successes on galaxy to cluster scale, and agree with the larger scale observations. A CDM adiabatic model dominated by vacuum energy with $\Omega_{vac} = 0.8$ and a CDM model with isocurvature initial conditions are shown to have extra large scale power over the standard adiabatic CDM model, though not enough to explain the observations. Further, the redshift of galaxy formation is too low in both models and attempting to bias galaxy formation to high peaks makes matters much worse.

1. ADIABATIC AND ISOCURVATURE CDM UNIVERSES

In inflationary cosmologies, the global density parameter Ω is almost exactly one and the density fluctuations are scale-invariant and Gaussian. There are two possible modes for the fluctuations: adiabatic and isocurvature. Adiabatic fluctuations arise from quantum zero point oscillations in the de Sitter vacuum of the inflaton scalar field which drives inflation. The case for an isocurvature mode is not as strong as for the adiabatic one. This mode can be generated by quantum zero point oscillations in a pseudo-Goldstone boson field such as the axion which is massless and dynamically unimportant during inflation. Density fluctuations in the axions then arise after the axion mass is generated at the chiral phase transition at $T \sim 200$ MeV. In this model, the axions would also form the the cold dark matter. Consideration of this model is somewhat pedagogical since it can be ruled out very convincingly on the basis of producing large-angle microwave background anisotropies in excess of those observed[1]. We denote by CDM the adiabatic model with $\Omega_X \approx 1$, $\Omega_B \ll \Omega_X$ and $h = 0.5$, where $h \equiv H_0/(100 \ km \ s^{-1} \ Mpc^{-1})$ and Ω_X is the density parameter of the cold dark matter. The isocurvature model with the same parameters is denoted by ISOC.

Since primordial nucleosynthesis abundances seem to constrain the baryon density parameter to be $\Omega_B < 0.1$, to satisfy $\Omega = 1$ dark matter should have a large non-baryonic component. Massive relics of the early universe can be classified as cold, warm or hot[2]; a substantial fraction of the dark matter could also be vacuum energy associated with a nonzero cosmological constant, $\Omega_{vac} \equiv \Lambda/(3H_0^2)$. The only hybrid model I consider here is VAC/C, with $\Omega_{vac} = 0.8$, $\Omega_X \approx= 0.2$, $\Omega_B \ll \Omega_X$. Such a small value for the vacuum energy density has little motivation from particle theories.

The shape of the density and velocity linear fluctuation spectra for adiabatic perturbations with cold dark matter depends only upon the combination $\Omega_X h^2$, provided $\Omega_B \ll \Omega_X$. The power spectrum for the density, $|\delta_X(k)|^2$, flattens from the natural initial Zeldovich-Harrison $n = 1$ slope, which remains on very large wavelengths, to $n \approx -1$ on cluster scales, to $n \approx -2$ on galactic scales, and to $n \approx -3$ on smaller scales. For $h = 0.5$ the power spectrum falls off too rapidly with increasing wavelength beyond cluster scales to explain the clustering properties of rich clusters and the large streaming velocities. Modification of the Zeldovich ($n = 1$) initial conditions by adding an $n = -1$ ramp between wavenumbers $k^{-1} = 5 - 300$ $h^{-1} Mpc$, then letting the ramp fall down to match the Zeldovich initial conditions again on large scales, does give enough power to explain the observations (§4). We denote this model with the same parameters as CDM by CDM+ϵ. There appears to be no compelling way within the inflationary paradigm to tune the onset of the ramp to begin at cluster scale.

Ideally, a complete inflationary model would tell us what the overall normalization amplitude of the fluctuation spectrum is. In the absence of such a theory, I choose a convenient parameterization of the normalization amplitude, the biasing factor b, defined so that mass traces light for $b = 1$ and mass is less clustered than light if $b > 1$. The technical definition of b relates a statistical average of mass density fluctuations to a volume average of fluctuations in the density of bright galaxies determined from the CfA redshift survey:

$$\langle \frac{\Delta M}{M}(< r = 10 \ h^{-1} Mpc) \frac{\delta\rho}{\rho}(0)\rangle \equiv b^{-2} \langle \frac{\Delta N_g}{N_g}(< r) \frac{\delta n_g}{n_g}(0)\rangle \approx b^{-2} 0.81. \qquad (1.1)$$

2.   COSMIC STRUCTURES AS PEAKS OF THE SMOOTHED DENSITY

To discuss the formation and clustering of objects of mass M, we smooth the mass density perturbation field $\delta(\vec{r}) = \delta\rho(\vec{r})/\rho$ on mass scale M; that is, we filter the density fluctuation spectrum on (comoving) length scale R_f. The relation between the mass scale and (Gaussian) filtering length is

$$M = 1.1 \times 10^{12} \ (R_f/Mpc)^3 \ h_{50}^2 \ M_\odot. \qquad (2.1)$$

Thus, a galactic scale smoothing would be $R_g \sim 1 \, Mpc$, $M \sim 10^{12} \, M_\odot$, and a rich cluster scale smoothing would be $R_{cl} \sim 10 \, Mpc$, $M \sim 10^{15} \, M_\odot$. For definiteness in the following, we choose $R_g = 0.35 \, h^{-1} Mpc$ and $R_{cl} = 5 \, h^{-1} Mpc$.

Moments of the filtered power spectrum on scale R_f reflect the *rms* level of the fluctuations in the density and velocity at time t, $\sigma_0(R_f, t)$ and $\sigma_v(R_f, t)$, respectively. The linear growth law is used to extrapolate these *rms* levels into the nonlinear regime; for the $\Omega_X \approx 1$ universes considered here, the redshift at which the *rms* fluctuations of scale R_f reach nonlinearity is $1 + z_{nl}(R_f) = \sigma_0(R_f, t_0)$, where t_0 is the present time.

I assume cosmic structures form at the smoothed peaks of the density field. The height of a peak is characterized by $\nu \equiv (\delta\rho/\rho)/\sigma_0$. The number density of peaks with heights in the range ν to $\nu + d\nu$, $\mathcal{N}_{pk}(\nu)d\nu$, can be determined using the mathematical theory of Gaussian random fields. The result is given in Bardeen *et al.*[3], hereafter BBKS. A selection function must be chosen for peaks which are to form a given class of cosmic objects. This could generally depend upon complex criteria associated with the structure of a peak. However, I will assume for the moment that it only depends upon ν, in which case it is $P(obj|\nu)$, the conditional probability that a peak of height ν forms an object of this class. The average number density of the objects is then

$$n_{obj} = \int P(obj|\nu)\mathcal{N}_{pk}(\nu)d\nu.$$

A straightforward example of such a selection function is the one adopted by Kaiser[4] and BBKS to describe rich clusters as $\sim 10^{15} \, M_\odot$ peaks which have collapsed by the present:

$$P(cl|\nu) = \vartheta(\nu - \nu_t), \quad \nu_t\sigma_0(R_{cl}, \, now) = f_c = 1.69. \tag{2.2}$$

Here ϑ denotes the Heaviside function and ν_t is the threshold. The value of the collapse factor f_c is somewhat arbitrary. $f_c = 1.69$ corresponds to a uniform spherical mass of cold particles in an $\Omega = 1$ universe having completely collapsed to a point in the center. Turn-around of the spherical perturbation occurs at 1.06, hence a value in between may be more appropriate. One way to normalize the spectrum would be to determine ν_t by matching equation (2.1) to the observed abundance of Abell clusters[5], $(55 \, h^{-1} Mpc)^{-3}$. For the CDM model this gives $\nu_t = 2.8$, with mean value $\langle\nu\rangle = 3.2$. Similar values hold for the other models and are given in Table 1. If we adopt $f_c = 1.69$, then we can use this to determine the biasing factor $b = 1.2$ in CDM, which is quite low.

In the simplest version of the biased galaxy formation theory presented by BBKS, *bright galaxies* - those used in the determination of the galaxy correlation function - are assumed to form from galaxy-scale peaks with selection function

$$P(gal|\nu) = (\nu/\nu_t)^q[1 + (\nu/\nu_t)^q]^{-1}, \tag{2.3}$$

where the power q must be high to ensure only high peaks are selected. $q = \infty$ corresponds to the sharp threshold discussed above. A fuzzier selection function with $q = 5$ is suggested in the physical model presented below. To get the threshold, we match the computed abundance to the density of bright galaxies determined from the CfA survey[6], $(4.6 \, h^{-1} Mpc)^{-3}$. For $q = 5$ and CDM we obtain $\nu_t = 3$ and $\langle\nu\rangle = 2.6$. The corresponding biasing factor determined using an expression for b given in §3.5 is 1.4. Values for other models are given in Table 1. (In Bond[7,8] I used the $q = 8$ results which have $b = 1.7$.) If only a fraction p of peaks above ν_t were to form Abell clusters, ν_t would be reduced: $b = 1.2$ if $p = 0.4$, in which case $\nu_t = 2.2$ and $\langle \nu \rangle = 2.4$. In this paper I adopt $b = 1.4$ for CDM. The fundamental difficulty with the biasing scenario is to explain how a global selection function similar to equation (2.3) can arise physically. Here I attempt to explain it in terms of local structural properties of the peaks.

3. A MECHANISM FOR BIASED GALAXY FORMATION

3.1 Gas Loss in Protogalactic Pancakes

The physics of the collapse of peaks is clearly crucial for determining the nature of the final object that forms. Here I argue that if there is collapse along one axis but continued expansion along the other two, the conditions are ripe for the formation of underluminous objects. This is due to gas loss from dwarf galaxy (dG) subclumps in the collapsing structure driven by the energy output of supernova explosions. This mechanism of gas loss from systems with low binding energy ($\equiv v_T^2/2$) has been discussed by Larson[9] and by Dekel and Silk[10]. A critical value for such loss is $v_T < 100\ km\ s^{-1}$ according to Dekel and Silk, where v_T is the three-dimensional rms velocity.

The picture I have begins with a primordial cloud a few times $10^{11}\ M_\odot$ in a CDM model. There will be clouds of all smaller scales embedded in it. Due to the Lin, Mestel and Shu[11] mechanism, the collapse will be asymmetric, preferentially collapsing along one axis to form a pancake. However, the pancake is decidedly inhomogeneous due to the substructure. These subunits may have already formed, or be in the process of forming, dwarf galaxies. Thus we should envisage a sheet-like cluster of dwarf galaxies. If expansion continues in the sheet, the gravitational binding energy holding the gas in the protogalactic medium decreases, making the supernova-driven gas ejection mechanism more likely to occur. Eventually, turnaround and collapse occurs along the other two axes. I will denote the time at which collapse occurs along the first (short) axis by t_1 and the time at which collapse occurs along the third (long) axis by t_3. If t_3/t_1 is high enough, there may be enough time to lose the gas from the dwarf galaxies and the inter-dG medium before collapse along all 3 axes makes the gravitational potential depth too large for the gas to escape. As I now show, constraining t_3/t_1 to be less than about 4 (e.g. collapse along axis 3 must have occurred before $z = 1$ if collapse along axis 1 occurred at $z = 4$) leads to a selection function with $q \approx 5$ and $\nu_t \approx 3$.

3.2 Virialized Properties of Spherical Top Hats

To get an idea of the potential well depths that peaks of height ν and smoothing radius R_f produce upon collapse and virialization, I use the simple spherical top hat model relating the following final virialized quantities (subscript V) to the redshift of collapse $1 + z_c = \nu\sigma_0(R_f, t_0)/1.69$ and the initial comoving top hat radius R_{th}:

$$n_{BV} = \frac{1000}{27} \bar{n}_{B0} (\nu\sigma_0(t_0))^3 = 4.2 \times 10^{-4} \Omega_B h^2 (\nu\sigma_0(t_0))^3\ cm^{-3},$$

$$v_T = H_0 R_{th} (\nu\sigma_0(t_0))^{1/2} = 100\ \frac{R_{th}}{h^{-1}Mpc} (\nu\sigma_0(t_0))^{1/2}\ km\ s^{-1}, \qquad (3.1)$$

$$T_V = \frac{m_N v_T^2}{3Y_T} = 4 \times 10^5 K\ Y_T^{-1} \left(\frac{R_{th}}{h^{-1}Mpc}\right)^2 \nu\sigma_0(t_0),$$

$$R_V = \frac{3}{10} R_{th} (\nu\sigma_0(t_0))^{-1},$$

$$M = \frac{4\pi}{3}\bar{\rho}(t_0)R_{th}^3 = 1.2 \times 10^{12}\left(\frac{R_{th}}{h^{-1}Mpc}\right)^3 M_\odot,$$

(3.2)

$$R_{th} = (9\pi/2)^{1/6}R_f = 1.56R_f.$$

Here m_N is the nucleon mass, $\bar{\rho}$ and $\bar{n}_{B0}$ are the average cosmological mass and baryon density, respectively, and Y_T is the number of particles per baryon in the gas (1.7 if it is fully ionized). I have dropped the explicit R_f dependence of σ_0 in this. Strictly speaking, top hat filtering rather than Gaussian filtering should be done to determine σ_0. The relation $R_{th} = 1.56R_f$ approximately takes this difference into account. $\Omega_X + \Omega_B \approx 1$ has been assumed.

For an isothermal sphere, the speed in circular orbits is related to v_T by $v_c = \sqrt{2}\,v_T$. In order to get the 220 $km\ s^{-1}$ circular speed of the Galaxy for our choice of R_g, we would require $\nu = 2 - 2.4$ for $b = 1.2 - 1.4$. Higher b values lead to higher ν to ensure the same collapse epoch. Our galaxy just about satisfies the criterion of being a bright galaxy. This crude argument demonstrates that deep potential wells only arise from high ν peaks. The empirical Faber-Jackson relation relating the luminosity of galaxies to the velocity dispersion $L \propto v_T^4$ demonstrates that if bright galaxies are selected by requiring that they be above some critical luminosity, this leads to a sharp threshold for ν, somewhere above 2. This is the physical basis for the *natural biasing* introduced by White *et al.*[12]: those objects, identified by the depth of their potential wells in N-body simulations of CDM models, which can conceivably become bright galaxies are already highly clustered in the initial conditions as in the simple biasing scenario. However, natural biasing does not give an explanation as to why galaxies of slightly lower velocity disperion are not bright. Further, there is no indication at present that such objects are clustered differently than the bright galaxies, as would be predicted.

The discussion above has been based upon a spherical collapse model. However, since real collapses are highly asymmetric, the final depth of the well will also depend upon the ratio t_3/t_1, with larger values giving significantly lower final dispersions. Thus requiring low t_3/t_1 selects objects that will be bright according to the empirical relation anyway. Taking gas loss into account may give an explanation for the effect.

Estimating the parameters of dwarf galaxies assuming $M_{dG} < 3 \times 10^{10}\ M_\odot$ ($R_{fdG} < 200\ h^{-1}Mpc$) and using the top hat model, we have, for $\nu = 2$ peaks, $v_T < 90\ km\ s^{-1}$ in CDM, so most of the subclumps would lose their gas if a critical velocity is indeed required for retention.

3.3   The Zeldovich Approximation

In the next two subsections, I determine the flow pattern of cold matter accreting onto a density peak in a Gaussian random field. To do this I utilize the Zeldovich solution for the trajectories of cold pressureless matter

$$x_i(\vec{r},t) = r_i - s_i(\vec{r},t), \quad s_i(\vec{r},t) = B(t)s_i(\vec{r}),$$

(3.3)

where r_i is the Lagrangian (unperturbed) position of the cold particle, Bs_i denotes the displacement field, and x_i is the Eulerian position. The Zeldovich approximation is the assumption of separability of the spatial and temporal parts of the displacement field in equation (3.3). For filtered spectra, the Zeldovich solution is exact in one dimension until caustic formation, and is quite accurate in three-dimensions until caustic formation, failing primarily by not giving dense enough clusters; and of course it reproduces perturbation theory to linear order. However, it is not clear how accurately it describes the flow pattern in a hierarchical picture since tidal forces generated by smaller scale collapsed structures could in principle dominate. I assume the validity of the Zeldovich approximation even in a hierarchy in the following.

The transformation from Lagrangian to Eulerian coordinates describes how the mass motion deforms the unperturbed FRW manifold. In Eulerian x-space, the peculiar velocity of the particles is $v_{pi} = (\partial x_i/\partial t)_r = -\dot{B}s_i(\vec{r})$, and the strain rate tensor (shear) is $(\dot{B}/B)e_{ij}$, where the strain tensor has components $e_{ij} = -\partial s_i(\vec{r},t)/\partial r_j$. The tidal tensor $\Phi_{,ij}$, the second derivative of the gravitaional potential, is given by $\Phi_{,ij} = -4\pi G\bar{\rho}a^2 e_{ij}$. The mass overdensity field is $\delta(r) == (1 + det((e_{ij})))^{-1} - 1$, which blows up when the the the largest eigenvalue of $1 + e$ reaches zero. I also introduce a random field for the trace of the strain tensor:

$$F(\vec{r},t) \equiv \nu(\vec{r},t)\sigma_0(t) \equiv -e_k^k(\vec{r},t).$$

F agrees with the relative overdensity field $\delta(\vec{r},t)$ in linear perturbation theory. The *rms* dispersion $\sigma_0(t)$ is then $\sigma_0^2(t) = \langle F^2(\vec{r},t)\rangle = [B(t)\sigma_0(t_0)/B_0]^2$. For homogeneous random fields, this is independent of $\vec{r}$, and in the Zeldovich approximation has a time dependence $B(t)$ given by the growth law for the relative overdensity in linear perturbation theory, the solution to

$$\ddot{B} + 2\frac{\dot{a}}{a}\dot{B} - 4\pi G\bar{\rho}B = 0, \quad B \to a \quad \text{as} \quad a \to 0.$$

For an Einstein de Sitter universe, $B = a$. In open models and models with a cosmological constant, $B_0 = B(t_0)$ is less than one.

Since σ_0 removes the B growth, it is natural to introduce variables with this factored out. Let the normalized eigenvalues of the strain tensor be defined by

$$e_{ij} = -\sigma_0(t)\sum_a \epsilon_a\, n_{ia}n_{ja}$$

where the triad of unit vectors $\vec{n}_a$ defines the principal axis reference frame, so n_{ia} are the components of the rotation matrix taking the i-axes into the principal ones. Introduce the traceless part of the shear (or strain) tensor $e'_{ij} = e_{ij} + (F/3)\delta_{ij}$; with eigenvalues $\epsilon'_a = \epsilon_a - \nu/3$. It is also convenient to define normalized variables in the i-frame:

$$v_R = -\frac{1}{2}(e'_{11} - e'_{33})/\sigma_0$$

$$w_R = \frac{3}{2}e'_{22}/\sigma_0, \tag{3.4}$$

$$A_1 = -e'_{23}/\sigma_0, \quad A_2 = -e'_{13}/\sigma_0, \quad A_3 = -e'_{12}/\sigma_0.$$

These 5 components plus ν define the 6 independent variables required to specify e_{ij}. In the principal axis system, v_R and w_R reduce to

$$v \equiv \nu e_v = (\epsilon_1 - \epsilon_3)/2$$

$$w \equiv \nu p_v = (\epsilon_1 - 2\epsilon_2 + \epsilon_3)/2. \tag{3.5}$$

The A_i degrees of freedom are absorbed into the rotation matrix components n_{ai} which can be parameterized by three Euler angles α, β, γ. The normalized measures of anisotropy in the strain tensor, e_v and p_v, are the analogue of the anisotropy measures in the mass quadrupole tensor introduced by BBKS, e, the ellipticity, and p, the prolaticity.

The relative total velocity of matter between a field point with position $\vec{r}$ and a field point with position $\vec{r}_p$ is

$$\Delta v_i \equiv v_i(\vec{r}) - v_i(\vec{r}_p) = Ha\Delta r_i + (H + H_B)e_{ij}(\vec{r}_p)a\Delta r_j, \quad \Delta r_j \equiv (r - r_p)_j, \quad (3.6)$$

where $H_B = \dot{B}/B$. $H_B = H$ in all models considered here but VAC/C, and at early times this relation still holds, deviating only when the vacuum energy comes to dominate; e.g., $H_B(t_0) = 0.41H_0$ at t_0 in VAC/C.

Equation (3.6) implies $\Delta v_{pa} = -H_B \epsilon_a \Delta r_a$ in the principal axis system. Thus, net accretion occurs onto the point r_p provided all three eigenvalues ϵ_a are positive. However, the mass grows provided $\nu > 0$, which is obvious in the linear regime. Peaks in the field F are points which maximize the mass accretion rate, and this remains valid in the nonlinear regime as long as the Zeldovich approximation works. This is the reason peaks are the fundamental points of interest for structure formation.

In the principal axis system, turnaround and complete collapse occur at redshift z_a and time t_a along axis a when $\epsilon_a \sigma_0(t_a) = 0.5$ and $= 1$, respectively. For models in which $B = a(t)$, we therefore have

$$e_v = \frac{z_1 - z_3}{6(1 + z_{av})}, \quad \nu\sigma_0(t_0)/3 = 1 + z_{av} \equiv 1 + \frac{z_1 + z_3}{2}. \quad (3.7)$$

Since

$$t_1/t_3 = \left(\frac{1 + z_3}{1 + z_1}\right)^{3/2} = \left(\frac{1 - 3e_v}{1 + 3e_v}\right)^{3/2}, \quad (3.8)$$

by restricting this ratio we are requiring that $e_v < e_{vm}$, with e_{vm} some fixed small number; for example, $t_1/t_3 < 4$ implies $e_{vm} = 0.14$.

3.4 The Distribution of Shear for Peaks

I now determine the distribution of shear eigenvalues for field points and, in an approximation, for peak points constrained to have height ν. For field points, it is straightforward to show that the variables ν, w_R, v_R and A_i are statistically independent and Gaussian distributed with zero means and dispersions 1, 1/5, 1/15 and 1/15, respectively. Rotation to principal axes and integration over the Euler angles defining their orientation gives the conditional probability distribution for e_v and p_v constrained by given ν and the restiction $|p_v| < e_v$, $e_v \geq 0$:

$$P(e_v, p_v|\nu)de_v dp_v = \frac{(15)^2\sqrt{5}}{2\sqrt{2\pi}} \, 2e_v(e_v^2 - p_v^2) \, e^{-5(\nu p_v)^2/2} e^{-15(\nu e_v)^2/2} \, \nu^5 de_v dp_v. \quad (3.9)$$

To get the distribution of the relative differences in collapse redshift along the 1 and 3 axes, we need the probability distribution for $e_v = v/\nu$, obtained by integrating equation (3.9). Since e_v is to be restricted to be less than the imposed maximum

e_{vm} which is small, the integration can be simplified. A further integration gives the probability that a field point will have e_v smaller than e_{vm},

$$P(< e_{vm}|\nu, \; field \; pt) = erf(\sqrt{10}\nu e_{vm})$$
$$+ \exp[-\frac{15}{2}(\nu e_{vm})^2] \; erf[(5/2)^{1/2}\nu e_{vm}] \; (1 - 15(\nu e_{vm})^2) \qquad (3.10)$$
$$- \frac{3}{2}\sqrt{10} \; \nu e_{vm} \; \exp[-10(\nu e_{vm})^2].$$

For small ν, the leading behaviour is

$$P(< e_{vm}|\nu, \; field \; pt) \sim (\frac{\nu}{2.71})^5 (6e_{vm})^5. \qquad (3.11)$$

Evaluation of the probability distribution $P(< e_{vm}|\nu, \; pk)$ for peaks is considerably more complicated, since integration over the local peak properties must occur. Here we discuss the form for a peak with the average properties without attempting to do such integrations. Analogous to the quantities ν, v, e_v, w and p_v introduced for the shear field, the corresponding quantities x y, e, z and p are introduced for the mass quadrupole tensor $F_{,ij}$ (BBKS). If we impose these local peak parameters as constraints along with the peak condition $\nabla F = 0$ for the conditional probability, then it turns out that $P(\nu, w_R, v_R, A_i|x, e, p, \; pk)$ is Gaussian distributed with ν, v_R, w_R and A_i statistically independent, having dispersions β, $\beta/5$, $\beta/15$, and $\beta/15$, respectively, with $\beta = 1 - \gamma^2$, but now having means γx, $\gamma x p$, $\gamma x e$ and 0, respectively. Here γ is a parameter determined from the fluctuation spectrum with values 0.60, 0.60, 0.50 and 0.53 for the models, in order, in Table I. The axes have been chosen to be the principal ones for the quadrupole tensor. Obtaining an equation for $P(v, w|\nu, \; pk)$ is now more difficult than for field points even if integration over the peak parameters is not performed. This is a result of the principal axes for the shear tensor not aligning with those for the quadrupole. The Euler angles rotating one set of axes into the other enter in the exponentials of the Gaussian, complicating further analytic simplifications. However, it turns out that for high ν, γx for peaks is $\sim \gamma^2\nu$, so the mean values are relatively small, suggesting as a first approximation that they can be ignored. This approximation requires detailed checking for lower values of ν, but I will adopt it as a hypothesis. In that case, the derivation for peaks follows that for field points, with the only modification being the presence of β:

$$P(< e_{vm}|\nu, pk) = P(< e_{vm}/(1 - \gamma^2)^{1/2}|\nu, \; field \; pt), \qquad (3.12)$$

an effective increase of the parameter e_{vm} that we are free to choose by only about 25%, using the values of γ quoted above. Thus I will adopt equation (3.10) as the selection function for galaxies.

If we identify the small ν behaviour of equation (3.10) with that of equation (2.3), we would have $q = 5$ and a relation between the threshold ν_t and the maximum e_{vm}:

$$\nu_t = 2.71(6e_{vm})^{-1} = 2.71\frac{[(t_3/t_1)_m^{2/3} + 1]}{2[(t_3/t_1)_m^{2/3} - 1]}, \qquad (3.13)$$

where $(t_3/t_1)_m$ is the maximum value for the collapse time ratio that will still allow bright galaxies to form. The $q = 5$ threshold function agrees with the form of equation (3.10) in the critical small ν regime up to $\sim 0.7\nu_t$, then is slightly higher up to $1.8\nu_t$, by which time both forms are nearly unity. These small deviations do not modify the central result: a selection function of form (3.10) will give all of the positive features from galaxy to cluster scale of the biased CDM model as computed by BBKS.

3.5 Galaxy Clustering in the Biased CDM model

The correlation function of galactic-scale peaks consists of two parts. Within the initial conditions themselves the peaks are *statistically clustered*: $\xi_{pk,pk}^{stat}(r) = A\psi(r)$, where $\psi(r) \equiv \xi_{\rho\rho}(r)/\sigma_0^2$ is the normalized density-density correlation function ($\psi(0) = 1$). The amplification factor A depends only upon ψ and ν_t. For the $q = 5$ case, A is asymptotically $A_\infty = 2.3$, ranging upward somewhat as ψ increases due to nonlinear effects in the dependence on ψ (BBKS). A more accurate treatment[13] includes dependences on derivatives of ψ, but these do not appreciably change the result over the interesting regime from $\sim 1 - 10\ h^{-1}Mpc$. It is quite remarkable that, for the adiabatic CDM model, $\xi_{pk,pk}^{stat}(r) = (r/r_{0,stat})^{-1.8}$ over this range, with $r_{0,stat} \approx 2.5\ h^{-1}Mpc$.

In addition to the statistical clustering, as the mass density field grows in amplitude, the peaks are carried along with the mass, giving an extra *dynamical* contribution to the correlation function. This term is difficult to treat generally except by N-body methods. A simple analytic model adopted by BBKS for the total peak-peak correlation function,

$$\xi_{pk,pk}^{stat+dyn}(r) \approx ([\xi_{pk,pk}^{stat}]^{1/2} + [\xi_{\rho\rho}]^{1/2})^2, \tag{3.14}$$

should be accurate in the linear regime, and seems to accord well with N-body studies provided the Universe is not too dynamically 'old' on galactic scales (to avoid the correlation function steepening precipitously with age).

The biasing factor b defined by eq.(1.1) is only slightly greater than the asymptotic value

$$b_\infty = A_\infty^{1/2}/\sigma_0 + 1 \tag{3.15}$$

for CDM. However, through equation (1.1), σ_0 depends upon b ($= 4.9/b$ in the model CDM). Solving for b with $A_\infty = 2.3$ gives $b = 1.45$ for $q = 5$, the value adopted in §2. We require that the galactic peak-peak correlation function (3.1) agree with the observed ξ_{gg}. The correlation function reaches unity at $r_0 = 5.7\ h^{-1}Mpc$, to be compared with the value $5.4 \pm 0.3\ h^{-1}Mpc$ of the CfA redshift survey[14]. Table 1 demonstrates how well this prescription works for the other models. A crude modification of ξ_{gg} for ISOC and VAC/C which uses the more correct biasing factors b_∞ listed in Table 1, which were determined self consistently from equation (3.5) since $\sigma_0 \propto b^{-1}$, is to multiply the values listed by 0.65 and 0.62, respectively. This brings $\xi_{gg}(5\ h^{-1}Mpc)$ into line, but at the expense of lowering the value for smaller r. This problem is a result of ψ being flatter than the required 1.8 slope, due to the spectrum's flatness. For CDM, ψ does have about the right slope, and $\xi_{pk,pk}^{stat}$ is even more closely a 1.8 power law.

3.6 Redshift of Galaxy Formation

The redshift when peaks with $\nu = \langle \nu \rangle$ collapse is given by

$$1 + z_g = \langle \nu \rangle \sigma_0(R_g)/f_c, \quad f_c \approx 1.69. \tag{3.16}$$

Hence, $z_g = 4.1$ for CDM. The parameters in Table 1 can be used to estimate z_g for the other models. Through the relation $1 + z \propto \nu$, the BBKS expression for $\mathcal{N}_{pk}(\nu)d\nu$, and the selection function, the galactic-scale peak collapse rate can be determined. For the fuzzy selection function used here for the bright galaxies, collapses would be spread out from $z \sim 6 - 3$. However, not without a strong constraint on e_{vm}, expansion along the 3-axis could continue until $z \sim 1$. These relatively low redshifts for galaxy formation could prove to be an embarassment for the theory if a great burst of energy accompanies collapse. Since smaller scale clouds within the protogalaxy would collapse first in this hierarchical model, the period of maximum energy release could be earlier than z_g. The rate of peak collapse (without a selection function being folded in) rises with decreasing z until its maximum around $z \sim 0.5$. These late collapsers would form underluminous low surface brightness objects since most would be highly asymmetric and possibly still expanding.

The ISOC and VAC/C models with biasing already have substantial statistical clustering for the peaks, as measured by the larger values of b required. With b_{∞} used for these two cases, unacceptably low redshifts of galaxy formation are obtained, $z_g = 0.7$ and 0.6, respectively. This problem is aggravated for sharper threshold functions, for then b_{∞} is higher. Even if light traces mass, $\sigma_0(R_g) = 2.6$, we again get 0.6 for the collapse redshift of rms fluctuations for ISOC. For VAC/C, the different fluctuation growth law B for vacuum-dominated universes gives complete collapse of the rms structures not occurring until $z \approx 1.1$. It is not clear how regions consisting of rms field points could actually form bright galaxies. The z_g problem can again be traced to the flatness of the spectrum between galaxy and cluster scale: ISOC has $n = -2.5$ on galaxy scale and CDM has $n = -2$.

4. CONSTRAINTS FROM LARGE SCALES

In this section, I use 3 indicators for large scale power, the cluster-cluster correlation function[5], ξ_{cc}, the cluster-galaxy correlation function[15], ξ_{cg}, and the large scale streaming velocities reported by Burstein et al.[16] and Collins et al.[17] to test the four $\Omega = 1$ CDM models considered here.

For rich clusters, the statistical clustering dominates over the dynamical contribution, which are included using equation (3.14). In Table 1, values of $\xi_{pk,pk}^{stat+dyn}$ for cluster-scale peaks selected according to equation (2.2), with ν_t chosen to agree with the cluster abundance, are shown for a few separations. $\xi_{pk,pk}^{stat}$ is evaluated in a linear approximation[13] to the full expression which is exact in the far field ($r > 4R_f \sim 20 \, h^{-1}Mpc$), and is not wildly off in the near field. The smoothing function chosen largely determines the shape in the near field anyway. Inclusion of the terms involving derivatives of ψ which were unimportant for the evaluation of ξ_{gg} are important for models such as CDM for which the spectral index n is steepening toward 1. CDM clearly does poorly beyond $r = 20 \, h^{-1}Mpc$. How seriously one views the failure of a model not having a correlation function which stays positive out to 100 or even out to 50 $h^{-1}Mpc$ depends upon one's attitude as to the reliability of the Abell catalogue. However, for ξ_{cc} to plunge as the CDM model indicates at the scale 25 $h^{-1}Mpc$ where we can see groups of clusters seems unlikely.

The cross-correlation function of the cluster-scale peaks with the mass density smoothed on cluster scale, $\xi_{pk,\rho}$, can also be evaluated. It depends not only upon ψ and its derivatives and the peak properties but also linearly on σ_0. The $\xi_{pk,\rho}$ results do not apply in the immediate neighbourhood of the cluster, for then the correlation function is measuring the collapse profile of the clusters, dynamics which is not included. To evaluate the cluster-galaxy correlation function, I assume the number density of galaxies at any point is a nonlinear function of the local smoothed mass density in the 'peak-background split' approximation of BBKS; in the far field $\xi_{pk,g} = b\xi_{pk,\rho}$, where

b is the biasing factor. This expression was used in Table 1. The ad hoc model with extra power, CDM+ϵ, and VAC/C both do reasonably well matching ξ_{cc} and ξ_{cg}.

There have been two recent reports that on very large scales, comparable to those probed by the cluster-cluster correlation function, the streaming velocities (bulk motions) of galaxies are $\sim 700 \ km \ s^{-1}$. Provided the filtering scale R_f associated with the observational selection is large enough so that the velocities on that scale are linear, the filtered linear fluctuation spectrum for the velocity can be used. The probability distribution for the streaming velocity is a Maxwellian distribution of speeds with 'temperature' per mass $\sigma_v(R_f)^2/3$. There is only a 5% probability of finding a bulk speed above $1.6\sigma_v$, and a 5% probability of finding one below $\sigma_v/2.97$. Table 1 gives the range of velocities interior to these values within which the speed lies 90% of the time for our 4 models. The Gaussian filtering radii are chosen to roughly correspond to the elliptical galaxy results[16] and the Rubin-Ford sample results of Collins $et \ al.$[17]. The models ISOC and VAC/C do not quite have large enough streaming velocities if mass traces light and are very far off with biasing. CDM is also a dismal failure, but its ad hoc modification CDM+ϵ is successful, though by design.

The CDM model with biasing is a success story from galaxy to cluster scale. The ISOC and VAC/C models do not fare well when biasing is included, having redshifts of galaxy formation which are too near to the present. Without biasing, the redshifts are still too small, since then rms fluctuations must collapse to produce galaxies. It is important to emphasize that every model in which galaxies form from peaks in the primordial density fluctuation field, as is expected, will have biasing as I discussed in detail in §3. A plausible physical mechanism was given for biasing which only depends upon local properties of the field and the Larson-Dekel-Silk model of gas loss from dG's; it does not require the exotic mechanisms to suppress galaxy formation at distant locations that have been discussed in the literature[18,19]. Whether the case against the CDM model with biasing on large scales is considered strong enough to rule it out depends upon the reader's assessment of the reliability of the data. It is certainly thought provoking that adding an ad hoc ramp onto the CDM spectrum from cluster scale to $\sim 300 \ h^{-1} Mpc$ gives a fluctuation spectrum which can satisfy all of the large scale tests discussed here, while still preserving the smaller scale successes. It is not clear how such a spectrum will arise however.

This work was supported by a Sloan Fellowship, a Canadian Institute for Advanced Research Fellowship and by NSERC.

TABLE 1: PARAMETERS FOR $\Omega = 1$ MODELS WITH BIASING

	CDM	CDM+ϵ	ISOC	VAC/C	OBS
b	1.44	1.44	1.44	1.44	
b_∞	1.45	1.50	2.18	2.33	
$\sigma_0(R_{dG} = 0.1)$	5.71	5.36	2.41	2.66	
$\sigma_0(R_g = 0.35)$	3.36	3.16	1.83	1.81	
ν_{tg}	3.05	3.05	2.48	2.74	
$\langle \nu \rangle_g$	2.58	2.58	2.38	2.46	
$A_{\infty g}^{1/2}$	1.51	1.51	1.43	1.49	
$\xi_{gg}(2.5)$	4.50	4.23	4.21	3.78	4.00
$\xi_{gg}(5)$	1.34	1.33	1.91	1.85	1.15
$\sigma_0(R_c = 5)$	0.49	0.51	0.51	0.52	
ν_{tc}	2.80	2.71	2.61	2.48	
$\langle \nu \rangle_c$	3.20	3.12	3.04	2.93	
$A_{\infty c}^{1/2}$	2.14	2.08	1.96	1.87	
$\xi_{cc}(20)$	0.17	0.92	0.76	1.38	1.49
$\xi_{cc}(25)$	0.09	0.73	0.45	0.98	1.00
$\xi_{cc}(30)$	0.03	0.55	0.27	0.70	0.72
$\xi_{cc}(50)$	-.01*	0.23	0.03	0.20	0.29
$\xi_{cc}(100)$	-.005*	0.06	-.003*	0.02	0.14
$\xi_{cg}(20)$	0.10	0.27	0.26	0.41	0.49
$\xi_{cg}(25)$	0.03	0.17	0.13	0.26	0.33
$\xi_{cg}(30)$	0.01	0.12	0.07	0.18	0.24
$v(R_f = 25)^{17)}$	60-286	240-1150	103-496	80-382	~700
$v(R_f = 40)^{18)}$	41-213	218-1048	77-371	66-317	970±300

For clarity of presentation, all models were chosen to have the same biasing factor, that appropriate to CDM. ISOC and VAC/C should have the larger b_∞ values for b to be consistent. This would lower ξ_{gg}, but give smaller values for σ_0 and the redshift of galaxy formation (equation 3.15). These models do better with $b = 1$, but it is not clear how this would be realized physically (§3). Equation (3.14) was used to compute ξ_{gg} and ξ_{gg}, except for the starred cases (*) which are negative. For these only the statistical clustering term was used. All r and R_f are in units of $h^{-1} Mpc$.

34

5. REFERENCES

1. Efstathiou, G. P. and Bond, J. R. : *Mon. Not. Roy. Astron. Soc.* **218**, 103 (1986).

2. Bond, J. R. and Szalay, A. S.: *Astrophys. J.* **277**, 443 (1983).

3. Bardeen, J. M., Bond, J. R., Kaiser, N. and Szalay, A. S. (BBKS): *Astrophys. J.* **304**, 15 (1986).

4. Kaiser, N. : *Astrophys. J. Lett.* **284**, L49 (1984).

5. Bahcall, N. and Soneira, R.: *Astrophys. J.* **270**, 70 (1983).

6. Davis, M. and Huchra, J.: *Astrophys. J.* **254**, 437 (1982).

7. Bond, J. R., "Large Scale Structure in Universes Dominated by Cold Dark Matter", in B. F. Madore and R. B. Tully, eds. *Galaxy Distances and Deviations from the Hubble Flow.* Reidel, Dordrecht (1986)

8. Bond, J. R., in S. Faber, ed. *Nearly Normal Galaxies from the Planck Era to the Present.* Springer (1987)

9. Larson, R.: *Mon. Not. Roy. Astron. Soc.* **169**, 229 (1974).

10. Dekel, A. S. and Silk, J.: *Astrophys. J.* **303**, 39 (1986).

11. Lin, C. C., Mestel, L. and Shu, F. H.: *Astrophys. J.* **142**, 1431 (1965).

12. White, S., Frenk, C., Davis, M. and Efstathiou, G. preprint (1986)

13. Bardeen, J. M., Bond, J. R., Jensen, L. G. and Szalay, A. S., in preparation.

14. Davis, M. and Peebles, P.J.E.: *Astrophys. J.* **267**, 465 (1983).

15. Seldner, M. and Peebles, P.J.E.: *Astrophys. J.* **215**, 703 (1977).

16. Burstein, D., Davies, R., Dressler, A., Faber, S., Lynden-Bell, D., Terlevich, R. and Wegner, G. in B. F. Madore and R. B. Tully, eds. *Galaxy Distances and Deviations from the Hubble Flow.* Reidel, Dordrecht (1986)

17. Collins, C. A., Josephs, R. D. and Robertson, N. A., *Nature* **320**, 506 (1986)

18. Rees, M. J.: *Mon. Not. Roy. Astron. Soc.* **213**, 75P (1985).

19. Silk, J.: *Astrophys. J.* **297**, 1 (1985).

Cosmic Strings, Galaxy Formation and Peculiar Velocities

Robert H Brandenberger
Department of Applied Mathematics and Theoretical Physics
University of Cambridge
Cambridge   CB3 9EW
UK

Abstract

In the first two thirds of this talk I explain what cosmic strings are and how they can lead to structures on cosmological scales such as galaxies and clusters of galaxies. In the final section I summarize some recent work on the peculiar large-scale streaming velocities predicted in the cosmic string theory of galaxy formation.

1.   What are Cosmic Strings?

Cosmic strings are linear topological defects which arise in some – but not all – gauge theories. They form lines of trapped energy density, and as such they are important in cosmology[1,2].

I will first discuss the basic ideas in terms of a simple toy model, a complex scalar field ϕ with a "Mexican hat" potential (Figure 1)

$$V(\phi) = \frac{\lambda}{4}\left\{|\phi|^2 - \sigma^2\right\}^2. \tag{1}$$

In particle physics, scalar fields with a potential similar to the above are used to break symmetries. Hence σ is called the scale of symmetry breaking. The values of ϕ which minimize the potential energy form a circle, the vacuum manifold M

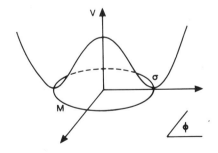

Fig. 1 Potential of the complex scalar field ϕ used to break the symmetry in the toy model for cosmic strings.

36

$$M = \left\{ \sigma e^{i\alpha} \mid 0 \leqslant \alpha \leqslant 2\pi \right\}. \tag{2}$$

The first point is to show that topological defects exist in this model. I will do so by constructing one. Consider a circle C_r of radius r in space (Figure 2) and fix the values of ϕ on this circle

$$\phi(r,\theta) = \sigma e^{i\theta} \tag{3}$$

This is a field configuration with "winding number" one, i. e. ϕ winds once round the vacuum manifold circle M as $\underline{x}$ winds one time around the circle in space. Given these boundary values of ϕ and assuming that ϕ is continuous, it is easy to show that ϕ cannot remain in M everywhere on the disk D (Figure 2) bounded by C_r. We assume the contrary and derive a contradiction. If $\phi(\underline{x}) \in M$ for all $\underline{x} \in D$ then we can consider $\phi(\underline{x})$ on smaller circles $C_{r'}$, with $r' < r$. The winding number (an integer) must remain 1 by continuity, and as $r' \to 0$ we reach the contradiction with single-valuedness of ϕ. Thus there must be a point $\underline{z} \in D$ for which $\phi(\underline{z}) \in M$. The above argument can easily be refined to show that there exists a point $\underline{z} \in D$ for which $\phi(\underline{z}) = 0$. At this point there is a lot of potential energy stored in the scalar field configuration.

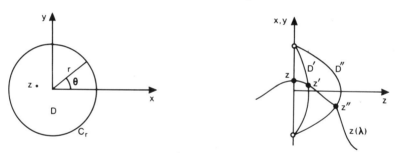

Fig. 2 The geometry for the existence argument given in the text. On the circle C_r boundary conditions for the scalar field ϕ are imposed which correspond to a field configuration with nonvanishing winding number. Then there exists at least one point $\underline{z}$ on the disk D bounded by C_r for which $\phi(\underline{z}) = 0$.

Fig. 3 The sequence of deformations $D(\lambda)$ of the disk D. On each disk there is a point $\underline{z}(\lambda)$ with $\phi(\underline{z}(\lambda)) = 0$. The $\underline{z}(\lambda)$ form the center line of the string.

Next I argue that the defects in this model are linear (and hence called strings). Consider a one parameter family of deformations $D(\lambda)$ of the disk D with fixed boundary $\partial D(\lambda) = C_r$. By the above argument, for each λ there is a point $\underline{z}(\lambda) \in D(\lambda)$ with $\phi(\underline{z}(\lambda)) = 0$. The spatial gradient energy in the class of field configurations with winding number 1 on C_r is minimized if the defect points $\underline{z}(\lambda)$ form a line (Figure 3). The line cannot have ends, since otherwise D could be deformed around the end and we would reach a contradiction with the argument in the previous paragraph. Cosmic strings are therefore either closed loops or infinite.

The example considered above is a model with global strings arising from a spontaneously broken global symmetry. There are several features which make global strings less attractive to cosmology than local strings, strings which arise when a local gauge symmetry is broken. There are Goldstone bosons and thus long range forces between widely separated segments of the string. The classical analysis of the evolution of the net-work of strings (sketched below) does not apply. The mass per unit length of a global string in flat space is infinite, and global strings decay rapidly by Goldstone boson emission[3]. To avoid having to deal with these problems, I will only discuss local strings, topological defects with a well defined width $\sigma^{-1}\lambda^{-1/2}$ and mass per unit length $\mu \sim \sigma^2$.[4].

2. Formation and Evolution of Strings

Not all gauge theories admit cosmic strings[5]. In order for them to do so, the vacuum manifold $\mathbf{M}$ must be non simply connected, i.e. $\Pi_1(\mathbf{M}) \neq 1$, where Π_1 is the first homotopy group. On the other hand, if we consider a gauge theory which admits cosmic strings, then defects will inevitably arise in the early universe. The following argument is due to Kibble[6].

From statistical mechanics it is well known that the state of a system in thermal equilibrium minimizes the (temperature dependent) free energy. Applied to a quantum field theory, this means that the field configuration at any temperature T is obtained by minimizing the finite temperature effective potential $V_T(\phi)$. The temperature dependence of $V_T(\phi)$ is sketched in Figure 4. For $T > T_c$, $\phi(x) = 0$ minimizes $V_T(\phi)$, for $T \ll T_c$ the energetically favored states are $\phi(x) \in \mathbf{M}$. Thus as the universe cools down to a temperature $T < T_c$ there will be a phase tran-sition at which $\phi(\underline{x})$ will change from 0 to a value in $\mathbf{M}$.

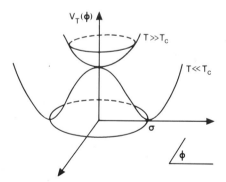

Fig. 4 One loop finite temperature effective potential of ϕ in the toy model for cosmic strings.

Any phase transition is characterized by a critical temperature T_c and a correlation length ξ. The significance of the latter is as follows : given two points with $|\Delta\underline{x}| > \xi$ there is no relation between the values of ϕ taken on at the two points. By causality we know that the correlation length must be smaller than the horizon. In a radiation dominated universe therefore

$$\xi(t) \leqslant 2t. \qquad (4)$$

On scales larger than ξ the phase of $\phi(\underline{x})$ will be random. Hence at the time of the phase transition a network of cosmic strings with mean curvature radius and separation ξ will form.

The evolution of the network of cosmic strings is nontrivial. If it were trivial, the energy density in strings would rapidly come to dominate the energy density in the universe, a cosmological disaster. To see this, consider a fixed comoving volume V containing a string. If the evolution were trivial, then the energy density in string would scale as $a(t)^{-2}$ compared to the energy density in radiation which scales as $a(t)^{-4}$. (In this talk I will only consider flat Friedmann-Robertson-Walker (FRW) universes with a scale factor $a(t)$.)

Two mechanisms contribute to decrease the energy in the cosmic string network. The first is production of string loops. On the large scales, the infinite strings look more like a random walk then like a straight line. They also typically have large translational velocities. Hence self-intersections of a string are common. Shellard[7] has numerically studied string intersections and concluded that unless the relative velocity exceeds 0.9c the strings intercommute. Thus, an infinite string can split off a loop. The loop radius R at the time of production t will typically be of the order $\xi(t)$ (Figure 5).

Fig. 5 Formation of a loop of radius R by a self intersection of an infinite string with curvature radius $\xi(t)$.

Once formed, loops oscillate and retain constant size in physical coordinates. Any oscillating mass will produce gravitational radiation. The power P of gravitational radiation from cosmic strings has been estimated analytically and numerically[8]. The result is

$$P = \gamma G \mu^2 \qquad (5)$$

where G is the gravitational constant and $\gamma \sim 10^2$ is a numerical constant.

The two mechanisms combine to give a way in which energy in strings is converted into radiation. In detailed numerical simulations, Albrecht and Turok[9] find that the correlation length $\xi(t)$ always remains proportional to the horizon t, i.e. the network of infinite strings is scale invariant. Kibble[10] has argued analytically that $\xi(t) \sim t$ is a stable fixed point under the dynamical equations of motion. The energy density in infinite strings then scales as radiation

$$\rho_\infty(t) \sim \frac{\mu t}{t^3} \sim \mu t^{-2} \sim a(t)^{-4} \qquad (6)$$

(in the radiation dominated FRW phase).

The scaling solution yields a simple formula for the number density $n(R,t)dR$ of loops of radius between R and $R + dR$. About one loop of radius t is produced at time t per horizon volume per expansion time. Hence

$$n(R,t = R) = \nu R^{-4}. \qquad (7)$$

Thereafter the number density simply redshifts. In the radiation dominated FRW phase hence

$$n(R,t) = \nu R^{-5/2} t^{-3/2}, \quad R \geqslant \gamma G\mu t \qquad (8)$$

The lower cutoff comes from gravitational radiation. For smaller R $n(R,t)$ is constant. Numerical simulations give $\nu \sim 10^{-2}$. The number density is independent of the mass per unit length μ, the reason being that the classical evolution of strings does not depend on μ.

3. <u>Formation of Structure</u>

Cosmic[1,11] string loops form massive seeds about which gas can accrete. I first want to compare this mechanism to the conventional one based on linear perturbations with random phases. In Figure 6 I sketch the energy density distribution in the two models at the time t_{eq} of equal matter and radiation, when pressure forces become unimportant and gas can begin to accrete onto seeds.

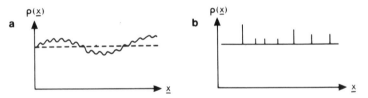

Fig. 6 Comparison of the energy density distribution at the time t_{eq} of equal matter and radiation for the conventional scenario of galaxy formation (a) and for the cosmic string theory (b). A rough sketch.

In the conventional model we have a superposition of plane wave perturbations with random phases. Their evolution can be described by the linearized Einstein equations. The amplitude of the perturbations grows slowly until $\delta\rho/\rho \sim 1$. Then the overdense regions collapse. But collapse occurs very recently. No structures form early.

By contrast, in the cosmic string theory we have nonlinear seeds. In the language of plane wave perturbations this corresponds to nonrandom phases. The evolution is best described by a nonlinear collapse model, e.g. the spherical collapse model[12]. Objects begin to collapse at t_{eq}, so in this theory some structures form very early.

The models based on primordial linear fluctuations with random phases only work if the dark matter in the universe is cold (nonrelativistic). If the dark matter is hot (relativistic) then structures of galactic mass can only form by fragmentation of much larger objects since the primordial perturbations on small scales are washed out by free streaming[13]. But these fragmentation models do not work without violating the upper bounds on large scale perturbations from the absence of observable microwave background radiation anisotropies. The cosmic string theory, however, works for both hot and cold dark matter, since the seed loops are not destroyed by free streaming.

Small loops at t_{eq} give rise to galaxies, large loops to clusters of galaxies. This idea can be made more quantitative as follows. For each class of objects (galaxies, clusters, etc) a mean separation d of two objects is known. By demanding that the cosmic string theory give the correct number density of these objects, the radius $R(d)$ of the corresponding seed loop can be determined. If $n_{>R}(t)$ denotes the number density of loops of radius greater than R, then $R(d)$ is determined by

$$n_{>R(d)} d^3 = 1 \qquad (9)$$

Notice that this analysis is independent of μ, the only free parameter in the theory. μ does enter into the determination of the mass $M(R(d))$ which accretes onto a loop of radius $R(d)$. This two step calculation is sketched in the following diagram

$$d \longrightarrow R(d) \longrightarrow M(R(d))$$

cosmic string   Newtonian   (10)
model   gravity

independent of μ   depends on μ

To conclude this section I will mention 2 examples. The mean separation of galaxies (clusters of galaxies) is

$$d \sim 5h^{-1} Mpc \quad \left[d \sim 55h^{-1} Mpc \right]. \qquad (11)$$

Hence the radius of the seed loop for galaxies (clusters of galaxies) is[14]

$$R(d) \sim 5.10^{-3}(\Omega h)^2 t_{eq} \quad \left[R(d) \sim 0.7(\Omega h)^2 t_{eq} \right]. \qquad (12)$$

where h is the Hubble constant in units of $100 km\, s^{-1} Mpc^{-1}$. Note, however, that the theory predicts the existence of object with a continuous range of masses, and a definite number density distribution $n(M)$.

4. Predictions and Tests

The cosmic string theory makes a lot of definite predictions. Those involving only the distribution of objects, e. g. the correlation function of clusters, are independent of μ, those involving the mass of collapsed objects depend on μ, but on no other free parameters.

The first prediction is the scale invariance of the correlation function. According to the scaling solution, the network of infinite strings looks identical at any time t when measured in terms of the horizon distance. Thus the correlation function ξ_i of loops formed at time t_i should be independent of t_i when measured in distances scaled to t_i, or equivalently (since $d_i \sim t_i$) scaled to d_i, the mean separation of the loops, i.e.

$$\xi_i(r) = f\left[\frac{r}{d_i}\right] \qquad (13)$$

where f is some universal function whose precise form is yet to be determined. There is increasing observational evidence[15] that the correlation function is scale invariant. Observations of groups[16] of galaxies, clusters of different richness classes[17] and superclusters[18] are all consistent with

$$\xi_i\left[\frac{r}{d_i}\right] \sim 0.3\left[\frac{r}{d_i}\right]^{-1.8} . \qquad (14)$$

Using the numerical simulations by Albrecht and Turok, Turok[19] has determined the function f which is predicted by the cosmic string theory. The agreement with the observational results is good : both amplitude and slope match (Figure 7), both are independent of μ.

The total amount of matter $M(R)$ which accretes about a seed loop of radius R depends on μ. In fact μ can be determined by demanding that $M(R)$ agrees with the known mass of a cluster for the value of R which corresponds to clusters (see (12)). Accretion onto a seed mass can be described by the spherical collapse model. It gives the time evolution of the physical radius $r(t)$ of a mass shell surrounding the seed mass δM. In parametrized form the evolution is given by

$$r(\theta) = \frac{r_i}{2\delta_i}(1-\cos\theta) \qquad (15)$$

$$t(\theta) = \frac{3t_i}{4\delta_i^{3/2}}(\theta-\sin\theta) \qquad (16)$$

where r_i is the initial radius, t_i the time when collapse can start, and

$$\delta_i = \frac{\delta M}{M(r_i)} \qquad (17)$$

where $M(r_i)$ is the mass inside the shell. Using these equations it is not hard to determine $G\mu$[14,20]. The result for a universe with $\Omega = 1$ and cold dark matter is

$$G\mu \sim 2h^{-1}10^{-6} \qquad (18)$$

For hot dark matter $G\mu$ is slightly larger since collapse can only start later and thus a larger seed mass is required in order to accrete a fixed amount of mass.

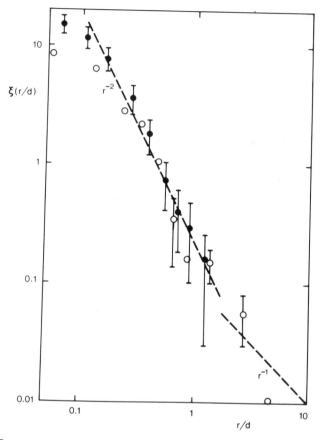

Fig. 7 Comparison of the correlation function of cosmic string loops (solid balls)[19] and of clusters of galaxies (open balls)[17]. The statistical errors of the results for string loops are indicated. The large statistical errors in the observational data points are shown only for a couple of data points. There may also be large systematic errors in the observational results.

The above value of $G\mu$ corresponds to a scale of symmetry breaking

$$\sigma \sim 10^{16} \text{GeV}, \qquad (19)$$

the scale of grand unification. It is a curious fact that the scale of unification has emerged with only astrophysical input data.

The above value for $G\mu$ can also be obtained by independent astrophysical considerations. The first is to do a calculation analogous to the previous one but for galaxies instead of clusters[14]. The second is to calculate the gravitational amplification of the galaxy two point correlation function. Since small loops are produced early gravitational forces have a long time to act and can produce significant changes in the initial correlation function. By demanding agreement with

$$\xi_g\left[\frac{r}{d_g}\right] \sim 4\xi_c\left[\frac{r}{d_c}\right] \tag{20}$$

where subscripts g and c refer to galaxies and clusters respectively, we can determine $G\mu$ [14]. The result agrees with (18).

All models which produce structures also give rise to anisotropies in the microwave background radiation and gravitational radiation. The microwave anisotropies in the cosmic string theory are safely below the current observational upper bounds. This is true for the scalar modes [21] and the tensor modes [22]. The most stringent bound on $G\mu$ comes from the stability of the millisecond pulsar [23].

5. Large Scale Peculiar Velocities

Recent observations [24] indicate that a large region of the universe of radius $60h^{-1}$Mpc has a net peculiar streaming velocity with respect to the microwave background of between 400 and 1000kms^{-1}. This compares to the velocity of 650kms^{-1} of our local group (a region of radius $5h^{-1}$Mpc) with respect to the microwave background. I should stress, however, that the validity of the above results is not yet generally accepted.

The cold dark matter model with linear adiabatic perturbations, random phases and the usual normalization of the power spectrum (no biasing) predicts a velocity of about 150kms^{-1} on the larger scale. Introducing biasing reduces the power spectrum and the velocities even further [25]. In hot dark matter models with random phases and with the normalization of the power spectrum chosen such that the first quasars can form at redshift 3, the large scale velocities would agree with the recent observations, but the mean velocity on the scale of the local group would be much too large (Figure 8). The main point is that it is impossible to reconcile the two velocity observations in a model with adiabatic perturbations, random phases and Zel'dovich spectrum, independent of amplitude.

Since the cosmic string theory is a model with nonrandom phases, it is interesting to estimate the scale dependence and amplitude of peculiar streaming velocities in this model.

The main result [26] is that in the cosmic string model the r.m.s. velocity is not a good measure of the typical velocity. Also, the distribution of velocities is not exponential about the r.m.s. velocity. Velocities larger than the r.m.s. velocity are more likely.

The typical peculiar streaming velocity is independent of scale λ for scales smaller than the horizon at t_{eq}, i.e. $25h^{-2}$Mpc. On larger scales the velocities decrease as λ^{-1}, like in the conventional models (Figure 8).

At present, the amplitude of the velocity spectrum can only be determined up to one order of magnitude. The velocity depends on $G\mu$, and the uncertainties in its determination are one source of the uncertainty in

44

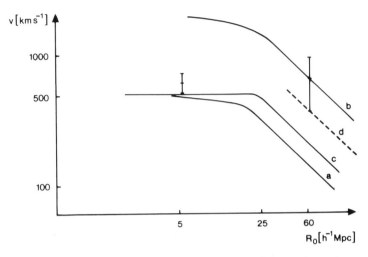

Fig. 8 Sketch of the dependence of the typical net streaming velocity δv
on the radius R_0 of the observation region
a) cold dark matter and random phases
b) hot dark matter and random phases
c) cosmic strings : typical velocities for $G\mu = 4.10^{-6}$
d) cosmic strings : peak velocities for $G\mu = 4.10^{-6}$

the velocity. In addition the velocities depend on parameters describing
the string loop distribution which are not well known.

I will now estimate the peak streaming velocities of an observation
region of radius R_0 due to a single cosmic string loop of radius R.
The peak velocity for given R is obtained when the separation d is of
the order R_0. The velocity will be estimated using the spherical collapse
model. We then maximize the velocity over R. The maximum value
$\delta v(R_0)$ is taken on when comoving scale $R(t)$ corresponding to R

$$R(t) = R\left[\frac{t}{R}\right]^{2/3} \qquad (21)$$

equals R_0, where t is the present time. The result on scales larger than
$2.5h^{-2}$ Mpc is

$$\delta v(R_0) \sim \beta G\mu \frac{t}{R_0} \sim 90(G\mu)_6\left[\frac{60h^{-1}\text{Mpc}}{R_0}\right] \text{kms}^{-1} \qquad (22)$$

where $(G\mu)_6$ is $G\mu$ in units of 10^{-6} and $\beta \sim 9$ is a parameter which
measures the average length of a loop with "radius" R.

We can similarly estimate the peak streaming velocity due to a single
wake. (For a discussion of wakes see e.g. Ref. 27.) The result
$\delta v_w(R_0)$ is almost identical to the result from single loops

$$\delta v_w(R_0) \sim 3\pi G\mu \left[\frac{t}{R_0}\right] \sim 100(G\mu)_6\left[\frac{60h^{-1}\text{Mpc}}{R_0}\right] \text{kms}^{-1}. \qquad (23)$$

The probability distribution of velocities due to loops scales as v^{-3}

$$P\left[R_0, g\delta v(R_0)\right] = g^{-3}P\left[R_0, \delta v(R_0)\right] \quad g \leqslant 1 \quad (24)$$

where $P(R_0, v)$ is the probability to observe a velocity v in an observation region of radius R_0.

The computation of the typical streaming velocities $v_T(R_0)$ is conceptually straightforward but technically somewhat tedious[26]. Using the value $v \sim 10^{-1}$ for primary loops we find

$$v_T(R_0) \sim 50(G\mu)_6 \left[\frac{60h^{-1}Mpc}{R_0}\right] kms^{-1} \quad R_0 > 25h^{-2}Mpc \quad (25)$$

and

$$v_T(R_0) \sim 150(G\mu)_6 \, kms^{-1} \quad R_0 < 25h^{-2}Mpc. \quad (26)$$

The large scale peak streaming velocities for the value $(G\mu)_6 = 2$ obtained with cold dark matter are comparable to the velocities obtained for cold dark matter models with random phases. They are too small to match the recent observations. With hot dark matter, the cosmic string theory gives large velocities, since $G\mu$ is larger. For $G\mu \sim 5$ it seems possible to achieve consistency with both the velocity measurements on scales of $5h^{-1}Mpc$ and $60h^{-1}Mpc$.

6. Conclusions

Cosmic strings provide a mechanism for producing structures in the universe which differs significantly from the models with primordial linear adiabatic fluctuations with random phases. In the cosmic string theory structures form by accretion onto seed loops. The theory predicts the existence of structures with a wide range of masses and with a scale invariant correlation function. The correlation function is independent of any free parameters in the theory and agrees both in amplitude and in shape with the observed correlation function of clusters of galaxies. Several independent astrophysical calculations give the same mass per unit length in strings (the only free parameter of the model). Such a mass would be induced by phase transitions at the grand unification scale.

Preliminary calculations indicate that the peculiar streaming velocities in the cosmic string theory have a different scale dependence than in the usual hot and cold dark matter models. On scales smaller than $25h^{-2}Mpc$ the typical peculiar velocities are constant. There is a value of $G\mu$ for which the velocities are marginally consistent with both the observed velocities on the scale of the local group and on a scale of $60h^{-1}Mpc$.

References

1. Ya. Zel'dovich, Mon. Not. R. Astr. Soc. 192, 663 (1980);
   A. Vilenkin, Phys. Rev. Lett. 46, 1169 (1981).

2. For a recent review see
   A. Vilenkin, Phys. Rep. 121, 263 (1985).

3. R. Davis, Phys. Rev. D32, 3172 (1985).

4. H. Nielsen and P. Olesen, Nucl. Phys. B61, 45 (1973).

5. D. Olive and N. Turok, Phys. Lett. 117B, 193 (1982).

6. T. Kibble, J. Phys. A9, 1387 (1976).

7. P. Shellard, Nucl. Phys. B. In press (1986).

8. T. Vachaspati and A. Vilenkin, Phys. Rev. D31, 3052 (1985);
   N. Turok, Nucl. Phys. B242, 520 (1984).

9. A. Albrecht and N. Turok, Phys. Rev. Lett. 54, 1868 (1985);
   A. Albrecht and N. Turok. In preparation (1986).

10. T. Kibble, Nucl. Phys. B252, 277 (1985);
    D. Bennett, Phys. Rev. D33, 872 (1986).

11. N. Turok, Phys. Lett. 126B, 437 (1983);
    A. Vilenkin and Q. Shafi, Phys. Rev. Lett. 51, 1716 (1984);
    J. Silk and A. Vilenkin, Phys. Rev. Lett. 53, 1700 (1984).

12. J. Peebles, The Large-Scale Structure of the Universe
    (Princeton Univ. Press, Princeton, 1980).

13. J. Bond, G. Efstathiou and J. Silk, Phys. Rev. Lett. 45, 1980 (1980).

14. N. Turok and R. Brandenberger, Phys. Rev. D33, 2175 (1986).

15. A. Szalay and D. Schramm, Nature 314, 718 (1985).

16. S. Shechtman, Ap. J. Suppl. 57, 77 (1985).

17. N. Bahcall and R. Soneira, Ap. J. 270, 20 (1983);
    A. Klypin and A. Kopylov, Soviet Astr. Letters 9, 41 (1983).

18. N. Bahcall and W. Burgott, Ap. J. (Letters) 300, L35 (1986).

19. N. Turok, Phys. Rev. Lett. 55, 1801 (1985).

20. H. Sato, Prog. Theor. Phys. 75, 1342 (1985).

21. R. Brandenberger and N. Turok, Phys. Rev. D33, 2182 (1986);
    J. Traschen, N. Turok and R. Brandenberger, Phys. Rev. D34, 919 (1986).

22. R. Brandenberger, A. Albrecht and N. Turok, Nucl. Phys. B277, 605 (1986).

23. C. Hogan and M. Rees, Nature 311, 109 (1984).

24. C. Collins, R. Joseph and N. Robertson, Nature 320, 506 (1986); D. Burstein et al., "Elliptical galaxies and nonuniformities in the Hubble flow" in "The extragalactic distance scale and deviations from the Hubble flow", ed. B. Madore (Reidel, Dordrecht 1986).

25. N. Vittorio, R. Juszkiewicz and M. Davis, "Large scale velocity fields as a test of cosmological models", Berkeley preprint 1986; J. Bond, "Large scale structure in universes dominated by cold dark matter", CITA preprint 1986.

26. R. Brandenberger, N. Kaiser, P. Shellard and N. Turok, DAMTP preprint (1986).

27. J. Silk and A. Vilenkin, in Ref. 11; T. Vachaspati, Phys. Rev. Lett. in press (1986); A. Stebbins, R. Brandenberger, S. Veeraraghavan, J. Silk and N. Turok, Berkeley preprint (1986).

GALAXY FORMATION WITH BARYONIC INFALL:
*Implications for Galaxy Dynamics, Decaying
Dark Matter and Dark Matter Detection*

Ricardo A. Flores[†]

Department of Physics, Brandeis University
Waltham, MA. 02254, U.S.A.
and
Theory Division, CERN
CH-1211, Geneva 23, Switzerland

ABSTRACT

The structure and properties of protogalaxies have been extensively studied in the gravitational instability model. In order to relate these protostructures to real galaxies, however, one must quantify the effect of the dissipational infall of its visible matter on the mass distribution of its dark matter (DM). We describe a simple analytic model to do this, that has been extensively checked by numerical simulations. We discuss the implications of the model for the dynamics of disk galaxies, cosmologies with decaying DM and detection of DM with superheated superconducting colloid (ssc) detectors.

1. INTRODUCTION

Galaxies have played a key role in the history of modern cosmology ever since the discovery by Hubble[1] that the Andromeda nebula was another island universe, much like our own Milky Way. As the smallest free standing objects in the universe, their dynamics has led to the discovery of the expansion of the universe and of the DM that seems to pervade all of space and is the dominant component of the mass in the universe. Their spatial distribution has led to the discovery of groups, clusters and superclusters of galaxies, in an ever larger hierarchy of structure in the universe. All of this has placed them at the forefront of research in cosmology.

An expanding, homogeneous universe (the Big Bang cosmology) is a solution to Einstein's equations and can account for the relative abundance of light elements and the temperature and nature of the microwave background. The universe at present, however, is highly inhomogeneous. Most modern theories of the origin of structure in the universe assume that the structure we see today, from the smallest galaxies to the largest superclusters, grew from gravitationally unstable, small density inhomogeneities. A special feature of an expanding universe is that it is unstable against the growth of small density perturbations: slightly overdense, bound regions first participate of the general expansion but later slow down to maximum expansion and subsequently collapse. Since any density inhomogeneity on cosmologically interesting scales would have been outside of its horizon (Hubble radius) at early times, a stumbling block for the model in the original Big Bang cosmology was that the perturbations would have had to be present *ab initio* and not set by causal processes. Inflation naturally circumvents this problem, since perturbations on microphysics scales are pushed outside of their horizon by the inflationary phase.

A great challenge to theory, of course, is to explain the great variety of structure that is observed. A promising model of formation of structure is a flat universe dominated by cold DM (CDM). The model assumes that the DM that is gravitationally dominant on all scales larger than galaxy cores is "cold" at early times: i.e., by definition the DM is non−relativistic very early on, so that perturbations on all scales larger than even the smallest galaxies are preserved (by contrast, "hot" DM (HDM) becomes non−relativistic much later and perturbations on all scales up to at least the largest galaxies are erased by free−streaming; models in which the gravitationally dominant form of matter is HDM appear to be ruled out[2]). The model can account for the structure seen on a large range of scales[3], provided that galaxies form **only** in unusually high peaks of the density field[4] (smoothed on galaxy scales). There is no reason other than simplicity to expect galaxies to be distributed like the underlying DM (i.e. that the average fluctuation on a given scale in, say, a galaxy count equals that of the mass on the same scale) and, in fact, several mechanisms[5] could be resposible for a biasing of galaxies toward high density peaks, including the gravitational biasing: fluctuations of the same size will turn arround at different times depending on the local density contrast. This latter effect might even be strong enough to provide most of the biasing needed[6].

Much effort has been devoted to the study of protogalaxies in this model[7] and in the gravitational instability model in general[8], in order to determine their structure and properties, but a first step needed to relate these protostructures to real galaxies is to quantify the effect of the dissipational infall of the visible matter on the mass distribution of the dark halo. A key difference between the DM and the visible matter (which I'll also refer to as "baryons") is that the latter can **dissipate** energy as electromagnetic radiation (while fully ionized, via bremsstrahlung, collisional exitation and radiative recombination). The DM is most likely to be dissipationless (halos are much less centraly condensed than the baryons) and, therefore, can only collapse by about a factor of 2 in radius from maximum expansion, while the baryons can sink deeply in the potential well of the halo via dissipation to form the compact luminous cores we see today: the visible galaxies. We describe here an analytic model[9,10] to calculate the radial redistribution of the DM caused by baryonic infall. Two regularities of the rotation curves of spiral galaxies can be readily explained by the model. First, the rotation curves of spirals are flat and featureless **across the optical boundary**, with no feature that indicates the different distributions of baryons and DM[11−13]. In the analytic model of infall[9] the gravitational coupling between these two mass components suffices to flaten the rotation curves across the optical boundary, provided that three conditions are satisfied: a) the angular momentum of the protogalaxies, which controls the amount of infall of the visible matter, is that expected in the tidal torque theory (which automatically operates in the CDM model, although the angular momentum is not very sensitive to the slope of the power spectrum[14]); b) the fraction by mass of dissipational material F is ~ 0.1, as seems to be the case observationally[15]; and c) collapsed protogalaxies have large core radii before baryonic infall occurs, as is found to be the case for CDM[7] as a result of the slope of the power spectrum on galaxy scales[8]. A second regularity is the existence of families of rotation curves[11]: rising, flat and falling profiles are seen and galaxies fill a narrow range of slopes around zero for their rotation profiles in the outer parts. Such systematics naturally arise in the infall model[10] as a result of the dispersion in angular momentum and core radii of protogalaxies, although a quantitative comparison with all the data is precluded until photometry of all galaxies in the sample becomes available.

This paper is organized as follows: in section 2 we describe the analytic model. Section 3 discusses numerical simulations of infall that relax many of the assumptions of the analytic model; one finds that the model works well indeed. Finally, section 4 discusses first the implications of the infall model for galaxy dynamics, such as those discussed above. Then we discuss the implications of the model for cosmologies with decaying DM, which have been proposed as a means to explain how the value of the mean density of the universe inferred from the dynamics of galaxies and clusters, about 20% of the critical density needed for a flat universe, could be compatible with a flat universe. We close with a remark relevant for proposed ssc DM detectors.

2. ANALYTIC MODEL OF INFALL

The dissipational collapse of the visible matter inside a protogalaxy will undoubtedly be very messy, with shocks and star formation affecting the gas in a complicated manner. However, observational constraints indicate that the mass fraction of dissipational (visible) matter in galaxies is small, $F \lesssim 0.2$[15]; since its coupling to the DM is likely to be purely gravitational, one can attempt a zeroth–order approximation to calculate the effect of the baryonic infall on the distribution of the DM by assuming that the baryons fall in toward the center to a prescribed distribution, thus slightly perturbing the mass inside a given radius. One can then expect the DM to move **adiabatically** inwards.

Blumenthal *et. al.* have studied in detail[9,10] a simple analytic model based on adiabatic invariance, extensively checked by numerical simulations, for calculating the radial redistribution of the DM of a protogalaxy when its baryonic matter falls in toward the center. Recall that for a particle moving in a periodic orbit $\oint pdq$ is an adiabatic invariant (*i.e.* its time derivative vanishes), where p is the canonical conjugate momentum of the coordinate q. Thus, for particles moving in circular orbits about a spherically symmmetric distribution of mass and $p = $ angular momentum, $rM(r)$ is the adiabatic invariant, provided that $M(r)$, the mass inside the orbital radius r, changes slowly compared with the orbital time.

Consider, then, a spherically symmetric protogalaxy of radius R that consists of a mass fraction $F \ll 1$ of baryons and $1 - F$ of DM. Assume that the two components are well mixed initially (i.e. that the ratio of their densities is F throughout the protogalaxy), as is likely to be the case since the DM would dominate the gravitational potential, and that no baryons fall inside the truncation radius R from beyond R. This is a rough approximation to the final shape of an expanding protogalaxy, which will cut itself out of the general expansion with some characteristic size, although there is no reason to expect it to be exactly spherical. In the absence of dissipation, R might represent the virial radius of the protogalaxy after violent relaxation.

Since there is more phase space for nearly circular orbits than for nearly radial orbits, we shall make the approximation that the DM particles move in circular orbits about the protogalaxy center (although for purely radial orbits the form of the adiabatic invariant is the same, $r_{max}M(r_{max}) = constant$, provided that M(r) varies in a self–similar fashion) with almost randomly oriented angular momentum vector (we assume that the protogalaxy has a small initial angular momentum). Thus, as the baryons fall in to the center to a final mass distribution $M_b(r)$, a

DM particle initially at radius r_i will move in to a radius $r < r_i$. The adiabatic invariant for such a particle implies that

$$r\left[M_b(r) + M_{DM}(r)\right] = r_i\,M_i(r_i),\qquad(1)$$

where $M_i(r_i)$ is the initial total mass distribution and $M_{DM}(r)$ is the final distribution of DM. If one assumes that the orbits of the halo particles do not cross, then

$$M_{DM}(r) = (1 - F)M_i(r_i)\qquad(2)$$

and equations (1) and (2) can be used to calculate the final radial mass distribution of the DM once $M_i(r_i)$ and $M_b(r)$ are given. If the dissipational mass fraction $F \ll 1$, then for a DM particle not too near the center of the protogalaxy, the mass interior to its orbit will change by a small fraction in one orbital period, even if dissipation occurs rapidly. Thus, adiabatic invariance is expected to be a good approximation for all but the innermost DM particles.

In order to use the adiabatic invariance equation (1), the initial mass distribution $M_i(r_i)$ must be assumed to be an equilibrium configuration. We shall assume that the initial protogalaxy relaxes to an isothermal sphere with core radius

$$a \equiv 3v^2/4\pi G\rho_o,\qquad(3)$$

where v is the one–dimensional velocity dispersion and ρ_o is the central density. As we first emphasized[9], large values of the core radius, $a/R \sim 0.4$, are nedeed for baryonic infall to produce mass distributions with flat rotation curves. Such large values would arise naturally if protogalaxies had a substantial amount of kinetic energy at maximum expansion, as is found to be the case for a universe dominated by cold DM[7] as a result of the slope of the power spectrum on galaxy scales: a/R is rather sensitive to the power spectrum[8], with $a/R \sim 0.1(0.4, 1.0)$ for a $n = -1(-2, -3)$ power law perturbation spectrum of the form $|\delta_k|^2 = k^n$ (δ_k is the Fourier transform of the density fluctuation field).

The final baryonic mass distribution, $M_b(r)$, would depend upon galaxy type. Elliptical galaxies (and the bulges of spirals as well) have surface brightnesses $\mu(r)$ that are well fit by a de Vaucouleurs $r^{1/4}$ profile $\mu(r) = \mu_e exp[1 - (r/r_e)^{1/4}]$, which would be the mass surface density as well under the usual assumption of constant M/L for the visible matter; they are triaxial bodies, show little rotation and seem to be supported by anisotropic velocity dispersions. Thus, it is usually assumed that in the case of elliptical galaxies baryonic infall is stoped by star formation. It has been shown[16] that the dissipational collapse of a lump of matter that has a negligible amount of angular momentum does indeed lead to a $r^{1/4}$ density profile over a large range of radii as a result of the energy loss. Unfortunately, the amount of dissipative infall in this case, r_e/R, would depend on the density threshold that sets star formation on, which is not known beyond order–of–magnitud estimates[17]. Spiral galaxies, however, are very cold and flat rotating disks. Thus, baryonic infall is most likely stoped by conservation of angular momentum in this case and, therefore, the final radial mass distribution of the baryons is constrained by the initial angular momentum of the protogalaxy. The final mass distribution of the baryonic material will be assumed to be

$$M_b(r) = M_{disk}(r) = M_{disk}\left[1 - (1 + r/b)exp(-r/b)\right],\qquad(4)$$

which is the radial mass distribution of a thin disk of mass M_{disk} whose surface density

$$\mu(r) = \left(M_{disk}/2\pi b^2\right)exp(-r/b) \qquad (5)$$

decreases exponentially (as the observed surface brightness) with scale length b. Numerical simulations[18] indicate that the flattening of the distribution of baryons does not appreciably distort the halo from spherical symmetry. Thus, it seems a reasonable approximation to assume the baryonic mass distribution to be spherical when using equation (1).

The relative scale size of the disk, b/R, can be related to the initial dimensionless angular momentum of the protogalaxy, $\lambda \equiv J|E|^{1/2}/GM^{5/2}$, assuming that the formation of the disk involved no transfer of angular momentum between the visible and the dark matter[10,19]. Here J, E and M are the total angular momentum, energy and mass of the protogalaxy. Simple theoretical arguments show[20] that if protogalaxies receive most of their angular momentum through tidal torques, then one expects the population of protogalaxies to have a mean λ of ~ 0.07. This has been confirmed in N$-$body simulations[14,21], which give $<\lambda> \sim 0.07$ with a width for the distribution of $\Delta\lambda \sim 0.03$. Using adiabatic invariance, equation (1), it is straightforward to calculate b/R as a function of λ^{10}. For a thin exponential disk without a halo, roughly half of the angular momentum is contained by the outside 25% of the mass. Thus, the adiabatic invariance assumption should be quite good.

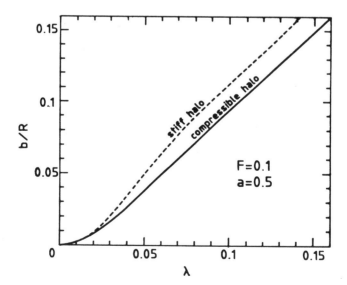

Figure 1: Relative scale size of the disk, b/R, versus the dimensionless angular momentum λ for $F = 0.1$ and a/R = 0.5. The solid (dashed) line represents the result when the DM does (does not) respond to baryonic infall.

Figure 1 shows how b/R is related to λ. If the halo is assumed to be stiff[19] (i.e. the halo remains with the distribution $M_i(r_i)$ without responding to the infall of the baryons) one obtains the dashed line. However, if the halo is allowed to respond one finds more infall (i.e. smaller b/R) at any given λ. This is because the DM halo **contracts** as a result of baryonic infall, with the ratio of final to initial DM mass inside a fixed radius, $M_{DM}^f(r)/M_{DM}^i(r)$, increasing for smaller radii (by definition of the truncation radius R, $M_{DM}^f(R)/M_{DM}^i(R) = 1$). Thus, for a given value of b/R, a stiff halo ($M_{DM}^f(r)/M_{DM}^i(r) = 1$ at all radii) produces a smaller rotational velocity for the disk at all radii and gives, accordingly, a smaller disk angular momentum. The relation between b/R and λ is also sensitive to the values of a and F; e.g. b/R is $\sim 25\%$ smaller (bigger) at $\lambda = 0.07$ for a^{-1} and F a factor of 2 bigger (smaller). See ref. 10.

The theoretical rotation curves of the model can also be calculated using the adiabatic invariant equation. Figure 2 shows the rotational velocity as a function of distance from the center, r/b, for typical values of F, a/R and λ. As expected, if the halo is not allowed to respond to the baryonic infall, a feature arises in the rotation curve at the "Holmberg" radius $R_H \equiv 4.5b$ (R_H is defined as the radius at which the surface brightness falls to 25th mag/arcsec^2, which for spirals of typical central surface brightness corresponds to $\sim 4.5b$), indicating the different distribution of the two components. However, as first emphasized in ref. 9, when the gravitational coupling between the two components is included, very flat rotation curves arise for $F \sim 0.1$, $\lambda \sim 0.07$ and a/R ~ 0.4. The morphology of the theoretical rotation curves in the region of R_H is varied, as a result of the dispersion in λ and a/R, and similar to that of the observations. This will be discussed in section 4.

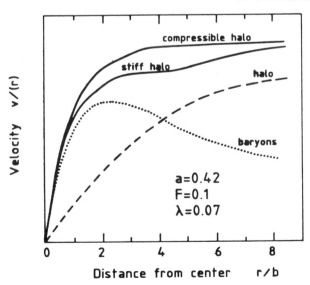

Figure 2: Rotation curves of the analytic model: $v(r)$ is the rotational velocity, in the plane of the disk, needed by a particle to remain in circular orbit at distance r/b from the center. The upper (lower) solid line is the result when the DM does (does not) respond to baryonic infall. The dotted (dashed) line is the contribution of the disk (initial halo) to the velocity.

3. NUMERICAL MODELS OF INFALL

The analytic model just described assumes circular or radial orbits, spherical symmetry, an equilibrium starting configuration, no shell crossing and does not include expansion. These assumptions can be relaxed and more general cases can be studied using N−body simulations. We carried out extensive numerical work to test the analytic model using a dissipative N−body code based on the code developed by Aarseth[22]. This work is discussed in detail in ref. 9; here we briefly outline its main features and show some of the results. One finds that the analytic model works well indeed.

The simulations follow a large number of particles (typically 1000) that interact via a Newtonian potential, softened on small scales in order to suppress spurious two−body relaxation effects due to the relatively small number of bodies in the simulation (the softening can also be interpreted as the presence of soft "clouds" instead of particles). Integration time steps were chosen so that the fractional change of total energy was less than 1% over a whole simulation, and the integration continued for several dynamical times after the system reached approximate equilibrium. The system was chosen to have total mass $M = 10^{12} M_\odot$ and to consist of (typically) 500 dissipationless particles (the DM) and 500 dissipational particles (the baryons). The masses were chosen so as to have F in the range $0.05 \leq F \leq 0.3$.

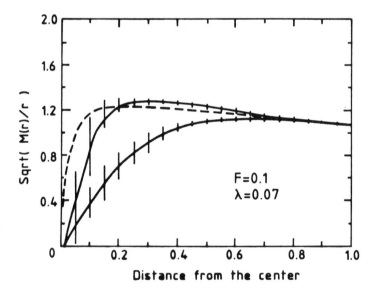

Figure 3: "Rotation" curves of a numerical model. The y−axis is $(M(r)/r)^{1/2}$, where $M(r)$ is, at fixed r, averaged over several dynamical times after the system reached approximate equilibrium. The bars are one standard deviation fluctuation levels. The x−axis is the distance from the center in units of the radius that contains 80% of the total mass (edge effects probably affect the results of the numerical models outside this radius). The dotted line is the result of the analytic model for $(M(r)/r)^{1/2}$

In simulations with dissipation, the baryonic particles were assumed to have a collisional cross section σ, which was kept constant throughout a simulation. A collision occurred if two particles came within a distance $(\sigma/\pi)^{1/2}$ of each other, in which case they were merged to form a single particle located at their center of mass, moving with its velocity and mass. Their relative energy was thus dissipated (σ controls the rate of baryonic infall; for larger σ the baryons fall in faster). We chose σ such that the total energy loss rate due to inelastic collisions equals the physical cooling rate (assuming fully ionized H + He). For more discussion see ref. 9.

An example of such simulations is summarized in Fig. 3, where the results of a "Hubble start" model are presented. The model starts with an expanding, homogeneous sphere with pure Hubble flow, except for a small amount of rigid body rotation with $\lambda = 0.07$; the fraction of baryons is $F = 0.1$. At the start time the Hubble constant is chosen so that the protogalaxy expands by a factor of ~ 2.5 before collapse. The lower solid line is the resulting "velocity" profile, $(M(r)/r)^{1/2}$, when no dissipation takes place. The upper solid line is the result for the same simulation when the baryons are allowed to "dissipate". The dashed line is the result of the analytic model for $(M(r)/r)^{1/2}$. The initial distribution needed for the analytic model was taken to be the final configuration of the dissipationless run (lower solid curve). The disagreement between the two models for $r \lesssim 0.15$ is due to the different final baryon distribution: for the analytic model, equation (4) was used; had we taken $M_b(r)$ from the numerical simulation, the two curves would coincide for $r \lesssim 0.1$[9]. The deviation from the numerical model for $r \gtrsim 0.2$ is due to the fact that the analytic model assumes that baryonic infall occurs after dynamical relaxation, while in the numerical simulations both processes occur simultaneously, therefore equation (2) is not exactly verified. The small core radius of the final distribution of the dissipationless run is a result of the unrealistic power spectrum used ($n = 0$) to distribute the particles inside the expanding protogalaxy. Larger core radii would have resulted for $n \simeq -2$[8].

The results are fairly insensitive to σ and agree fairly well with those of the analytic model for $F \lesssim 0.2$. Thus, the analytic model seems quite appropriate.

4. DISCUSSION

4.1 Implications For Galaxy Dynamics

As discussed in the introduction, the rotation curves of spiral galaxies exhibit remarkable regularities that certainly demand an explanation in any theory of galaxy formation. First, it has been pointed out by several authors[11-13] that the rotation curves of spirals are flat and featureless across the optical boundary, where the rotation of the gas and stars changes from being dominated by the baryons (inside R_H) to being dominated by the DM (outside R_H); it is as if there were a "conspiracy"[13] between the luminous and the dark matter to end up distributed just so as to erase any feature that might arise in the rotation curve that would indicate the different distribution of the baryons and the DM. The dissipative infall of the baryons compresses DM halos, shrinking the core radius of the initial isothermal distribution of the DM to much smaller values after infall[9] and leaving the DM with a radial density distribution that no longer resembles that of an isothermal sphere but is instead much more centrally peaked. Such compression leads to mass distributions with very flat and featureless rotation curves, as shown in Fig. 2, provided that a) the initial core radius is large, $a/R \sim 0.4$, b) the mass fraction of baryons is small, $F \sim 0.1$, and c) the initial angular momentum $\lambda \sim 0.07$. Each of these values is independently motivated: protogalaxies with core radii $a/R \sim 0.4$ naturally arise in a CDM universe[7] as a result of the slope of the power spectrum on galaxy scales[8] and they would also have $\lambda \sim 0.07$, although this value is not very sensitive to the slope of the power spectrum[14]; observational constraints, on the other hand, constrain F to the range $0.05 \lesssim F \lesssim 0.2$ and place it near 0.1 in rough order of magnitude[15]. As Fig. 2 shows, a feature does arise in the rotation curve in the absence of compression.

It has been noted also[11−13] that there are roughly equal amounts of disk and halo matter interior to the optical radius of a galaxy (which approximately corresponds to R_H). This is also the case in the analytic model: the ratio of halo to disk masses at R_H is indeed close to unity for $0.04 \lesssim \lambda \lesssim 0.1$ and all values of the core radius if $F \simeq 0.1$[10].

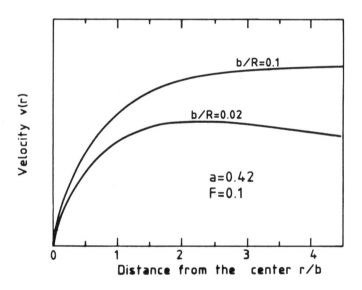

Figure 4: Rotation curves of the analytic model inside the "'Holmberg" radius, $R_H \equiv 4.5b$, for $F = 0.1$ and $a/R = 0.42$. The axes have the meaning of those of Fig. 2. The relative scale size $b/R = 0.1$ (0.02) corresponds to $\lambda \sim 0.1$ (0.04). The normalization corresponds to galaxies of the same scale size and total mass.

Another regularity that has been noted in the rotation of spirals is the existence of families of rotation curves[11]: rising, flat and falling profiles are seen and galaxies fill a narrow range of slopes around zero for their rotation curves in the outer parts. Such rich morphology of rotation curves is not at all unexpected in the infall model as a result of the dispersion in λ and a/R: although protogalaxies are likely to have a universal F (if most of the DM in the universe is non-baryonic; a variety of arguments[23] suggest that this is indeed the case, the weakest argument being against the DM being in objects of planetary mass), they are certainly not expected to have a universal λ or a/R, and the dispersion about their mean gets reflected in the theoretical rotation curves. Figure 4 shows the limiting rotation curves for the extremal values of λ and $F = 0.1$ and $a/R = 0.42$. We see that rising and falling rotation curves arise, reflecting the different ratios of halo to disk masses that result from the different degrees of infall, with intermediate rotation curves for $0.04 < \lambda < 0.1$. The effect of the dispersion in λ on the velocity profile is only **local**; the rotation curves are identical beyond $r/R \gtrsim 0.3$. The observed rotation curves differ significantly from these theoretical profiles for $r/b \lesssim 2$, but part of this difference can be accounted for by the presence of a bulge[10]. The dispersion in a/R (which is not well studied) has a similar effect, although the effect is **global** (rotation curves differ at all radii) and much less sensitive to a/R[10]. Furthermore, one finds that for $F \gtrsim 0.15$ only falling rotation profiles (i.e. $Deltav \equiv v(4.5b) - v(3.5b) < 0$) are possible for $0.04 \leq \lambda \leq 0.1$ and any value of a/R because a very centrally condensed mass distribution results in this case; likewise, for $F \lesssim 0.03$ only rising rotation curves (i.e. $Deltav > 0$) arise because of the absence of a centrally condensed mass distribution.

Observationally, there seems to be no correlation between Hubble type or total mass in visible matter and the form family[11], although it appears that an environmental correlation exists[24]: rising profiles are common among field galaxies while falling profiles arise in high density regions. Such would be the case in the infall model, since λ is expected to anti-correlate with local density[25].

Thus, the infall model discussed here can explain several regularities of the rotation of spiral galaxies for a region of parameter space that is motivated by entirely independent arguments. This is certainly motivating, and encourages further observational and theoretical study along the lines suggested by the model.

4.2 Implications For Decaying DM

The motivation for decaying DM is to reconcile the theoretically prefered Einstein-de Sitter universe, which has $\Omega = 1$ if no cosmological constant is present, with the observational constraint that on scales of up to $\sim 10h^{-1}Mpc$ the clustered matter amounts only to $\Omega_{cl} = 0.1 - 0.3$[15,26]. However, it should be noted that an underestimation of Ω by this much would be expected in a CDM universe with biased galaxy formation[3] and in the case of the evidence based on the infall of the Local Group toward the Virgo cluster it could also be the result of anisotropies in the infall flow[27]; furthermore, observations on very large scales[28] seem to be consistent with $\Omega = 1$, although they need to be confirmed.

In the simplest decaying DM models ("type I")[29] a heavy, relic elementary particle (e.g. a heavy neutrino of mass $\gtrsim O(1)keV$) first drives the formation of galaxies and clusters and subsequently decays non-radiatively into relativistic particles that still dominate today and provide a smooth, undetected background that contributes Ω_r to the total energy density of the universe. Thus, at present the non-relativistic matter contributes $\Omega_{nr} = \Omega_{cl}$ to the total energy density and $\Omega_r = 1 - \Omega_{nr} \sim 0.8$. Another possibility[30] ("type II" models) is that the universe has recently entered a second period of matter domination, after the energy of the relativistic decay products gets redshifted away and a primordial, stable non–relativistic species becomes dominant. In these models the decay of the unstable species unbinds large quantities of the stable species, causing them to stream away from clusters and larger structures, providing a smooth background of non–relativistic matter this time. Thus, $\Omega_{cl} \sim 0.2$ would only reflect the component of the DM clustered on small scales while the true $\Omega_{nr} \sim 0.8$. Using the analytic model of section 2, Flores et. al.[31] have considered the effect of decaying DM on the rotation curves of disk galaxies. They find that the existence of flat rotation curves out to large radii, as those seen in 21 cm observations, rule out type I models and that the type II models are significantly constrained. Reported very large scale flows[32] and large amplitud of the rich–cluster correlation function[33] would also rule out type I models[34], although not specifically but any pure cold– or hot–DM model of formation of structure. These observations, however, remain controversial[35].

The lifetime of the unstable species, τ, is constrained by several astrophysical considerations, most easily expressed in terms of the redshift corresponding to the epoch of decay, z_d $(1 + z_d \equiv s^{-1}(t = \tau)$, where $s(t)$ is the scale factor of the universe at time t, normalized to unity at the present). The isotropy of the microwave background[36], the virial equilibrium of the cores of rich clusters[37] and a model of gravitational lensing[38] all imply a similar upper bound $1 + z_d \lesssim 5$. The linear theory of our infall toward the Virgo cluster implies[37] $1 + z_d \gtrsim 20$, but this latter bound is exponentially sensitive to uncertainties in the observational parameters and is affected by the the asphericity[27] and non–linearity[39] of the infall, so a value as low as $1 + z_d \sim 5$ is not excluded. The upper bound, $1 + z_d \lesssim 5$, implies that the DM must decay on a time scale $\tau \gtrsim 10^9$ years, and since the mass fraction in visible matter at the present is small, $F \lesssim 0.2$[15], a DM particle's orbit about a protogalaxy is expected to change adiabatically during dissipation and decay. This allows one to calculate the effect of these processes on a galaxy halo with the model of section 2[31].

Theoretical rotation curves can be calculated as in section 2. The results are presented in Fig. 5, where F is the final fraction of baryons (relative to final total mass) after a fraction $1 - f$ (relative to initial total DM mass) of the DM decays. The two solid lines are the rotation curves for two values of f and typical values of F and b/R. The effect of the dispersion in λ is shown by the two dashed curves, which are rotation curves with the same values of F and f and λ at its upper (3) and lower (5) limit. Decreasing F produces less peaked rotation curves (as does increasing λ) as shown by the dotted line, which is a rotation curve with the same values of f and λ as those of curve (2) but $F = 0.05$.

The main feature of Fig. 5 is the decreasing velocity at large distances for small f. This can be readily understood: after the baryons have cooled to an exponential disk (dissipation and decay do not commute since λ is invariant under decay in the adiabatic approximation; we assume that by $t \sim \tau \gtrsim 10^9$ years the visible matter has dissipated and collapsed, since its dynamical and cooling times are shorter than τ), the DM decay does not significantly affect $v(r)$ at small radii because the baryons dominate there. At large radii, however, $v(r)$ decreases significantly as a result of the mass loss because the DM dominates there.

Extensive numerical work was carried out to confirm the validity of the analytic model, which was found to work quite well[31]. One can, therefore, use it to compute rotation curves and compare them to observations.

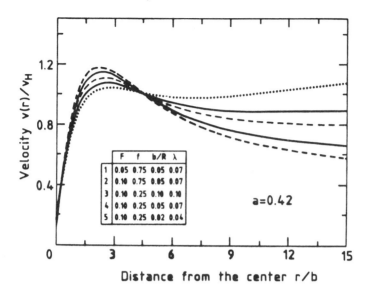

Figure 5: Theoretical rotation curves of models with DM decay. The axes have the meaning of those of Fig. 2 and $v_H \equiv v(r = R_H \equiv 4.5b)$.

Only a small fraction $\sim 20\%$ of the spiral galaxies in the Burstein and Rubin sample[11] are observed to have falling velocity profiles. For $f \lesssim 0.3$, however, Fig. 6 shows that all theoretical rotation curves fall $(\Delta v/v \equiv [v(4.5b) - v(3.5b)]/v(3.5b) < 0)$ inside the "optical" radius $R_H \equiv 4.5b$ for **all** values of a/R **even for F and λ^{-1} at their lowest possible values** (one can see in Fig. 5 that increasing F or decreasing λ only makes rotation curves fall more steeply around R_H). Furthermore, velocity profiles are observed to be flat or rising outside the optical radius as well[40] and this provides further constraints on f. Because we assume f to be the same for every protogalaxy (assuming that the DM left in halos is non−baryonic) one can use the data of individual galaxies to constrain it. For example, **no value** of a/R can yield a flat rotation curve out to $r = 11b$ (like that of NGC 3198[41]) unless $f \gtrsim 0.4$, nor can a profile rising out to $r \sim 8b$ (like that of NGC 3109[42]) be obtained if $f \lesssim 0.5$. These bounds are, again, for F and λ^{-1} taken at their minimum values.

Thus, the rotation curves of spiral galaxies require the fraction of stable DM $f \gtrsim 0.5$. In the type I model, the contribution to the energy density by relativistic particles required for $\Omega = 1$ amounts to $\Omega_r/\Omega_{nr} \sim 4$. This ratio can be related to f by[31]:

$$\Omega_r/\Omega_{nr} \leq f^{-1}(1 - f)(1 - F)(1 + z_d)^{-1}, \tag{6}$$

which even for a decay epoch as recent as $1 + z_d \sim 5$ requires $f \lesssim 0.05$, a factor of 10 smaller than the lower bound obtained here. Thus, with the assumptions

60

made[31], the type I model is ruled out. Only if a substantial fraction of the gas contracts after the DM decay might one lower the upper bound on f, although it is not clear that such a scenario can lead to flat rotation curves. In the type II models[30] $\Omega_r/\Omega_{nr} \sim 0.25$, which is marginally consistent with the upper bound of Eq. 6 even for a decay epoch as recent as $1 + z_d \sim 4$. However, an upper bound on f must come from the requirement that on large scales Ω be unity with Ω_{cl} being only ~ 0.2. We are trying to quantify this constraint by studying in detail the formation of clusters and large scale structure in the type II models[43].

4.3 Implications For DM Detection

Ssc detectors are being considered[44] as a possible detector of elementary particle DM. The idea is to have superconducting grains just below the critical temperature and embedded in magnetic fields, so that the energy deposited by a DM particle interacting in the grain would cause the grain to become normal, thus allowing the magnetic field to penetrate and producing a signal in the readout electronics as a result of the magnetic flux change. The number of readout loops is limited because of technical constraints and one might consider[45] maximizing the detection rate of each readout loop, which then scales with the rms velocity of the DM as v_{rms}^7 and is, therefore, extremely sensitive to it. In such case it is very important that the effect of baryonic infall on the DM halo be taken into account; we have seen that baryonic infall compresses the DM halo and the final halo profile deviates from isothermality in the inner parts. It is also important to calculate the effect on the DM of the formation of a real disk; numerical simulations[18] indicate that the velocity distribution of the halo is slightly perturbed from a maxwellian distribution.

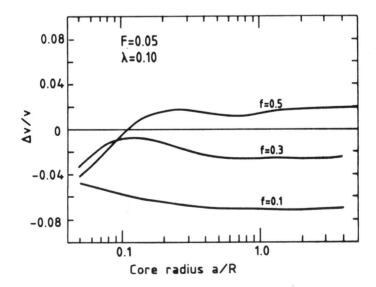

Figure 6: $\Delta v/v$ as a function of the core radius of the initial isothermal sphere for $F = 0.05$ and $b/R = 0.1$ (for reference, $\Delta v/v \sim -0.1$ for a self–gravitating exponential disk).

AKNOWLEDGEMENTS

I thank my collaborators G. Blumenthal, A. Dekel, S. Faber and J. Primack for letting me borrow material from our joint publications and for many fun discussions throughout our collaborations. I am also greatful to D. Seckel for useful conversations.

REFERENCES

† Present address: Theory Division, CERN.

1. Hubble, E., Ap. J. **69**, 103 (1929).

2. See *e.g.* Primack, J., to appear in the Proceedings of the 2nd ESO/CERN Symposium, G. Setti and L. Van Hove eds., March 1986.

3. White, S. D. M., Frenk, C. S., Davis, M. and Efstathiou, G., University of Arizona preprint, 1986.

4. Bardeen, J., in **Inner Space/ Outer Space**, E. W. Kolb *et. al.* eds., U. of Chicago Press 1986, p. 212. Kaiser, N. *ibid.*, p. 258.

5. For a review see *e.g.* Dekel, A., Comments Astron. Astrophys., in press (1986).

6. See the article by Frenk in this volume.

7. Frenk, C. S., White, S. D. M., Efstathiou, G. and Davis, M., Nature **317**, 595 (1985).

8. Quinn, P. J., Salmon, J. K., and Zurek, W. H., preprint, 1986.

9. Blumenthal, G., Faber, S., Flores, R. and Primack, J. Astrophys.J. **301**, 27 (1986).

10. Blumenthal, G., Faber, S., Flores, R. and Primack, J., UCSC preprint in preparation.

11. Burstein, D. and Rubin, V. C., Ap. J. **297**, 423 (1985).

12. Bahcall, J. N. and Casertano, S., Ap. J. **293**, L7 (1985).

13. Van Albada, T. S. and Sancisi, R., preprint 1986.

14. Efstathiou, G. and Barnes, J., in **Formation and Evolution of Galaxies and Large Scale Structures in the Universe**, J. Audouze and J. Tran Thanh Van eds., Reidel 1984, p. 361.

15. Blumenthal, G., Faber, S., Primack, J. and Rees, M., Nature **311**, 517 (1984) and references therein.

16. Calberg, R. G., Lake, G. and Norman, C. A., Ap. J. **300**, L1 (1986).

17. Rees, M. J. and Ostriker, J. P., M. N. R. A. S. **179**, 541 (1977).

18. Ho, K. S., private comunication.

19. Fall, S. M. and Efstathiou, G., M. N. R. A. S. **193**, 189 (1980).

20. Peebles, P. J. E., Ap. J. **155**, 393 (1969).

21. Efstathiou, G. and Jones, B. J. T., M. N. R. A. S. **186**, 133 (1979).

22. Aarseth, S. J. in **Multiple Times Scales**, J. U. Brackbill and B. I. Cohen eds., Academic Press 1985, p. 377.

23. Hegyi, D. J. and Olive, K. A., Ap. J. **303**, 56 (1986).

24. Burstein, D., Rubin, V. C., Ford, W. K. and Whitmore, B., Ap. J. **305**, L11 (1986).

25. Hoffman, Y., Ap. J. **301**, 65 (1986).

26. Peebles, P. J. E., Nature **321**, 27 (1986).

27. Davis, M. and Villumsen, J. V., Berkeley preprint, 1985.

28. Yahil, A., Walker, D. and Rowan-Robinson, M., Ap. J. **301**, L1 (1986). Loh, E. D. and Spillar, E. J., Ap. J. **307**, L1 (1986).

29. Turner, M. S., Steigman, G. and Krauss, L., Phys. Rev. Lett. **52**, 2090 (1984). Gelmini, G., Schramm, D. and Valle, J., Phys. Lett. **146B**, 311 (1984).

30. Olive, K. A., Seckel, D. and Vishniac, E., Ap. J. **292**, 1 (1985).

31. Flores, R., Blumenthal, G., Dekel, A. and Primack, J., Nature, in press.

32. Burstein, D. *et. al.* in **Galaxy Distances and Deviations from Universal Expansion**, B. F. Madore and R. B. Tully eds., Reidel 1986. Collins, C. A., Joseph, R. D. and Robertson, N. A., Nature **320**, 506 (1986).

33. Bahcall, N. and R. Soneira, Ap. J. **270**, 20 (1983).

34. Bond, J. R. in **Nearly Normal Galaxies: From the Plank Time to the Present**, S. Faber ed., Springer Verlag 1987.

35. Gunn, J. E., *ibid.* as ref. 35.

36. Silk, J. and Vittorio, N., Phys. Rev. Lett. **54**, 2269 (1985). Turner, M. S., Phys. Rev. Lett. **55**, 549 (1985). Kolb, E. W., Olive, K. A. and Vittorio, N., preprint, Fermilab-PUB-86/40-A, 1986.

37. Efstathiou, G., M. N. R. A. S. **213**, 29 (1985).

38. Dekel, A. and Piran, T., Weizmann preprint WIS-86/30-June Ph.

39. Hoffman, Y., Ap. J. **305**, L1 (1986).

40. Bosma, A., Astron. J. **86**, 1791 (1981); ibid. **86**, 1825 (1981).

41. van Albada, T. S., Bahcall, J. N., Begman, K. and Sancisi, R., Ap. J. **295** 305 (1985).

42. Carignan, C., Ap. J. **299**, 59 (1985).

43. Blumenthal, G., Dekel, A., Flores R. and Primack,J., in progress.

44. Drukier, A. K. and Stodolsky, L., Phys. Rev. **D30**, 2295 (1984).

45. Drukier, A. K., Freese, K. and Spergel, D. N., Phys. Rev. **D33**, 3495 (1986).

THE FORMATION OF STRUCTURE IN PARTICLE-DOMINATED COSMOLOGIES

Carlos S. Frenk

Physics Dept., University of Durham
Durham DH1 3LE England

ABSTRACT

The assumption that the "missing mass" consists of
weakly interacting elementary particles leads to
specific predictions for the formation of cosmic
structures. Cosmologies with different kinds of dark
matter are discussed. The most successful assumes that
the Universe is flat and is dominated by cold
collisionless relics. This model accounts for most
aspects of the large scale appearance of the galaxy
distribution and for the observed abundances and gross
properties of bound structures ranging from galaxy halos
to rich galaxy clusters.

1. INTRODUCTION

Particle physics has had a tremendous impact on classical cosmology
in recent years by proposing solutions to long-standing problems such as
the geometry of the Universe, the origin of density fluctuations and the
nature of the missing mass. In the inflationary model the cosmological
parameter Ω , defined as the ratio of the cosmic mean density divided by
the mean density of an Einstein-de Sitter universe with the same Hubble
constant, is predicted to be very close to 1 [1]. Quantum fluctuations
in the scalar field whose vacuum energy drives the inflation generate
causally connected curvature fluctuations with a scale-invariant,
constant curvature spectrum, known as the Harrison-Zel'dovich spectrum
2,3]. Weakly interacting elementary particles have been proposed as
candidates for the missing mass which is known to exist in the halos of
galaxies and in galaxy clusters and is thought to be responsible for the
inhomogeneous appearance of the Universe on large-scales. Candidate
particles range in mass from axions ($m \sim 10^{-5}$ ev), through neutrinos ($m \sim$
100 ev), to supersymmetric particles ($m > 1$ Gev). Although, with the
exception of the neutrino, none of these particles has yet been detected
in the laboratory, they are attractive to the cosmologist because they
lead to very specific models for the formation of cosmic structure that
can be tested against observations.

The theoretical framework within which galaxy formation and

clustering are normally considered is the gravitational instability of a Friedmann universe. Small density irregularities in a uniform expanding background are assumed to have been generated at an early epoch. At (re)combination the power spectrum of density fluctuations has the form

$$\left| \delta_k(t) \right|^2 \; = \; T(k,t) \left| \delta_k \right|_p^2 \tag{1}$$

where δ_k is the Fourier transform of the fluctuating density field, $\delta(\underline{r}) = \rho(\underline{r})/\bar{\rho}$, and k denotes spatial frequency. $\left| \delta k \right|_p$ is the primordial spectrum, usually assumed to be of the form $\delta_k^2 \propto k^n$ with random phases and T(k,t) is a transfer function that depends on the interactions between the various constituents of the Universe; inflation predicts n=1. (However, cosmic strings, are a possible mechanism to induce strong phase correlations at early times). Eventually self-gravity leads to the collapse of density inhomogeneities and their subsequent evolution is governed by non-linear processes which ultimately produce the galaxies, clusters and other structures characteristic of our Universe today.

If the Universe is dominated by weakly interacting elementary particles, the transfer function of equation (1) is determined by two different mechanisms. The first is Landau damping due to free-streaming of relativistic particles out of density enhancements; it produces an exponential cut-off in the power spectrum shortwards of a critical wavelength, $\lambda_c \propto m^{-1} \alpha (\Omega H^2)^{-1}$, where m is the particle mass and H=100h km s Mpc^{-1} is the Hubble constant [4]. The second mechanism is a reduced growth in the amplitude of matter fluctuations during the radiation era when the dominant photon-baryon fluid undergoes acoustic oscillations; it is known as the Mészáros effect [5,6] and produces a bend in the spectrum from the initial power-law of index n to n-4, at a characteristic scale $\lambda_c' \propto (\Omega H^2)^{-1}$ corresponding to the horizon size at the transition between matter and radiation dominance. Because the importance of free-streaming depends on the thermal velocity of the dark matter relative to the Hubble velocity at early times, elementary particle candidates have been classified into three broad classes, known as hot, warm and cold dark matter [7,8]. The fluctuation spectrum of hot dark matter is shaped by free-streaming, that of cold dark matter by the Mészáros effect and that of warm dark matter by both effects (fig. 1). The classic example of hot particles are neutrinos with mass of 30 ev for which the cutoff wavelength λ_c corresponds to 41 Mpc today [9,10]. Cold dark matter includes particles which have low velocities at early times either because they are very massive (eg photinos, gravitinos or Gev neutrinos) or because they are created in this state (the axion); less likely candidates for cold dark matter are quark nuggets [11] and primordial black holes [12]. Gravitinos and photinos have been suggested as candidates for warm dark matter [13,14] but currently it seems more likely that supersymmetric particles have a mass in excess of 1 Gev; for a hypothetical warm particle with mass ∿1 kev the free-streaming cutoff would occur at about a galactic scale.

It is clear from fig. 1 that the formation of structure will proceed very differently in universes dominated by hot, warm or cold dark matter. In a neutrino universe scales a few tens of megaparsecs across are the first to collapse; clusters and galaxies must subsequently fragment, probably from dense gaseous regions inside pancake-like superclusters. This kind of evolution produces a universe with considerable large-scale coherence. In a cold dark matter universe the fluctuation spectrum has power on all scales; subgalactic size objects are the first to collapse and larger structures must form by gravitational aggregation on a roughly hierarchical way. A similar kind of evolution on scales larger than individual galaxies is expected in a universe dominated by warm dark matter. In all these cases, up to a normalization constant and uncertainties in the cosmological parameters Ω and h, the "initial conditions" for the growth of structure are completely specified by the power spectra of figure 1 and the assumption of random phases. The qualitative non-linear features that I have just sketched can actually be calculated in considerable detail, at least in so far as gravity is involved.

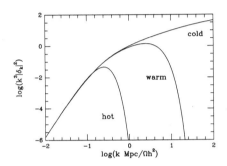

Figure 1. Power per decade in density fluctuations after (re)combination as a function of spatial frequency in universes dominated by hot, warm and cold dark matter. Adiabatic primordial fluctuations with the Harrison-Zel'dovich spectrum are assumed.

In the remainder of this paper I will discuss the later phases of evolution in hot and cold dark matter dominated universes. In its full generality the non-linear gravitational growth of density perturbations can only be followed by means of N-body simulations. In these, the trajectories of a large number of particles which can be thought of as a representative sample of phase-space elements, are calculated by solving the equations of motion in an expanding background universe. The simulations that I will describe below have been carried out over the past few years in collaboration with G. Efstathiou, M. Davis and S.D.M. White; a detailed account of their technical aspects and a dicussion of their range of applicability are given by Efstathiou et al [15] and previous reviews are given by Frenk [16] and White [17].

2. NEUTRINO UNIVERSES

At first sight neutrinos appear to be very attractive candidates for the missing mass. In contrast to all other elementary particle candidates they are known to exist and (as yet unconfirmed) measurements of a neutrino mass of about 30 ev were reported a few years ago [18,19]. Such a mass would resolve a number of important cosmological issues: it

66

implies that $\Omega=1$ and it singles out a preferred lengthscale in the early universe which corresponds roughly to the scale of present day galaxy superclusters. Unfortunately, when it was subjected to closer scrutiny this idea did not live up to such expectations mainly because the predicted large-scale structure does not agree with observations [20-22].

The dynamical evolution of a neutrino universe is dominated by the sharp cutoff in the fluctuation spectrum at a present day scale of a few tens of megaparsecs (fig. 1). This coherence length turns out to be much too large to be consistent with the observed clustering scale of galaxies. The neutrino distribution at late times obtained from the N-body simulations of White, Frenk and Davis [21] is shown in fig. 2 for an Einstein-de Sitter background universe. It has a marked filamentary appearance which reflects the coherence length of the initial power spectrum [20,23,24]. However, the filaments which join rich clusters and surround large low-density regions are a transient feature; soon after the epoch shown they break up into clumps and at a later time the distribution is completely dominated by a few large clusters containing most of the mass. In order to decide which epoch should be compared to the present, it is necessary to relate the neutrino distribution to the observed galaxy distribution. Galaxies can only form after the collapse of the first non-linear structures; in the simulations we identify the onset of galaxy formation with the time when 1 percent of the initial Lagrangian grid elements, used to specify the initial conditions, have collapsed to zero volume. The particles that have passed through these possible sites of galaxy formation are labelled as "galaxies" and are shown in fig. 2.

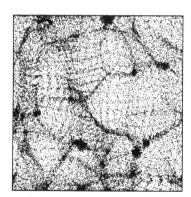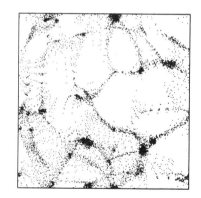

Figure 2. Projected distributions of neutrinos (left) and "galaxies" (right) in an N-body simulation of an Einstein-de Sitter neutrino-dominated universe in which galaxy formation began at $z=1.1$. The side of the box is 65 h^{-2}Mpc.

If the model time corresponding to this figure is identified with the present, galaxy formation would have occurred at a redshift of 1.1. This is far too recent, in conflict with the lower limit of 5 claimed for the redshift of galaxy formation [25] and with observations of quasars at redshifts of ~3. To avoid this problem we must identify a later, more clustered, epoch in the simulation with the present. A universe in which galaxy formation occured at a redshift of 2.5 is illustrated in fig. 3(b). To facilitate comparison with observations, this figure represents the model as seen by a hypothetical observer located at random in the computational volume who can only "see" particles selected with the same selection function of the Center for Astrophysics redshift (CfA) survey [26,27]. Both, neutrinos (filled circles) and "galaxies" (dots), are plotted. The corresponding projection of the CfA survey is shown in fig. 3(c). It is clear that the level of clustering of the neutrinos and even more so of the "galaxies" in the simulation far exceeds that of the galaxies in the Universe. The discrepancy is even worse in an open universe. A quantitative analysis of the simulations shows that the predicted distributions are in disagreement with the observations for any acceptable redshift of galaxy formation and for any light neutrino mass [21].

The arguments of the previous paragraph rely to some extent on the identification of galaxies in a simulation of the mass component. Although our identification algorithm is somewhat arbitrary, most conceivable physical mechanisms for galaxy formation would lead to a more clumpy distribution for the galaxies than for the neutrinos and would not therefore alter our conclusions. Nevertheless it might be argued that we know so little about galaxy formation that perhaps some way around these problems can be found. To allow for this possibility we considered the properties of neutrino clumps in our simulations, without making reference to the galaxy distribution [22]. At the present epoch these clumps contain about half the mass of the universe and their individual masses are so large that they would have no counterparts in the observable universe. Accretion of a small amount.of baryons onto these deep potential wells would turn them into strong x-ray sources which are not observed. Another idea that has been put forward to salvage the neutrino model is that some fraction of the neutrinos have a mass greater than 30 ev (and hence give rise to a smaller coherence length) but, in order to preserve a flat universe, they are assumed to have decayed non-radiatively into relativistic particles [28-31]. There is not much physical motivation for this idea and, moreover, Efstathiou [32] has argued that it is not possible to obtain bound galaxy clusters in this model without at the same time causing too large an acceleration of the Milky Way relative to the Virgo cluster of galaxies. Other arguments against massive neutrinos are the large streaming velocities predicted in this model [33] and the phase-space constraints which prevent neutrinos from condensing onto halos of dwarf spheroidals, if those halos do indeed exist [34]. Although it may be that a sufficiently contrived model can overcome all these objections, at the present time it appears extremely unlikely that massive neutrinos are the missing mass.

68

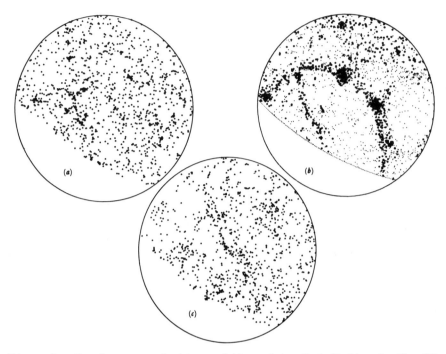

Figure 3. Equal area projections of the galaxy distribution in the CfA
catalogue (c) and of particle positions in simulations of an open
($\Omega=0.2$) cold dark matter universe (a) and an Einstein-de Sitter
neutrino universe (b). The outer circle represents Galactic latitude
+40°, while the empty regions lie at declinations below 0°.

3. COLD DARK MATTER UNIVERSES

3.1 Galaxy clustering

The most successful model of galaxy clustering so far assumes that
the dark matter is a cold collisionless relic [35,36]. The standard
post-recombination fluctuation spectrum for cold dark matter (CDM) tends
to the primordial power-law of index n=1 on large scales and gradually
tilts to an asymptotic power-law of index n=-3 on small scales (fig. 1).
Thus, in contrast with the neutrino universe, large scale structure in
this model builds up from subgalactic scales in a roughly hierarchical
way. Since the spectrum is very flat at high frequencies, a wide range
of scales becomes non-linear almost simultaneously; as a result galaxy
formation may be significantly influenced by environmental effects.
N-body simulations from these initial conditions require a large dynamic
range and a high resolution code is essential [15].

From simulations of the growth of structure on intermediate scales
($\sim$1-10 Mpc) we found two models which are in good agreement with the
galaxy data [36]. The first is an open model with $\Omega=0.2$ at present in

which the galaxies are assumed to be fair tracers of the mass. The
predicted galaxy distribution is displayed in fig. 3(a); its clustering
pattern, which has a trace of filamentary structure, is quite similar to
the observations (fig. 3c). This visual agreement is reflected in the
two-point correlation functions of the model and the observations which
have the same power-law slope and the same amplitude provided h=1.1 (For
separations between $\sim$ 0.1-10 Mpc, the galaxy two-point correlation
function is a power law of index $\sim$-1.8). The three-point correlation
function and the peculiar velocity distribution in the simulations do
not match the galaxy data in detail, but the differences are not great.
A more worrisome difficulty with this model is that it predicts
fluctuations in the microwave background on small angular scales which
marginally exceed the current upper limits [37-39]. In addition an open
model has the disadvantage that it conflicts with theoretical prejudice
and with the standard inflationary model, both of which demand that Ω be
equal to 1.

Not surprisingly, an Ω=1 model in which the galaxies are assumed to
be fair tracers of the mass does not match the galaxy data. This is a
reflection of the well established fact that matter clustered as
galaxies on scales of a few megaparsecs contributes only $\Omega \sim 0.2$ to
dynamical estimates of the mean cosmic density. (In the simulations this
discrepancy is reflected in an unacceptably low level of clustering and
unacceptably large peculiar velocities.) A flat universe can only be
reconciled with these measurements of Ω if the galaxies are more
strongly clumped than the mass; in this case only a fraction of the
total density contributes to virial estimates. Such a situation may
arise quite naturally in a cold dark matter universe where environmental
effects might be strong enough to restrict galaxy formation to regions
of unusually high density. Physical mechanisms that might inhibit galaxy
formation in low density regions have been proposed [40,41] . In
addition we have recently found that non-linear gravitational clustering
from cold dark matter initial conditions gives rise to a natural bias in
the population of galactic size objects which may account entirely for
the requirements of an Ω=1 universe [42,43]. This mechanism will be
dicussed in detail below. First I will review a simple ad hoc model
which has been extensively used to describe the statistics of galaxies
in a CDM universe. This model assumes that galaxies form only at those
peaks of the smooth linear density field that lie above a global
threshold [36,44-47]. It is specified by two parameters: the width, s,
of the smoothing filter (usually assumed to be Gaussian) and the height
above threshold, $\nu\sigma$, where σ is the rms density fluctuation. This
prescription selects a biased subset of the particles which follows the
overall clustering pattern of the underlying mass distribution but with
greatly enhanced amplitude [48].   Figure 4 shows the distributions of
dark matter (left-hand panel) and "galaxies" with ν=2.5 and s=0.5 h^{-1}
Mpc (right-hand panel) in one of the N-body simulations of Davis et al
[36].This bias brings a flat cold dark matter universe into agreement
with observations. Indeed, for h=0.5 the "galaxies" plotted in fig. 4
have approximately the same number density as bright galaxies, and
similar two- and three-point correlation functions and pair-weighted rms
peculiar velocities. In addition, the predicted fluctuations in the
microwave background are well within the allowed limits [37-39].

70

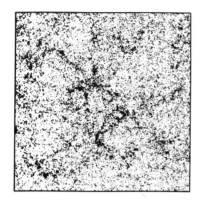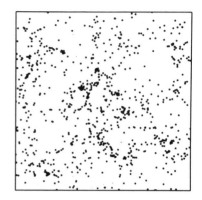

Figure 4. Projected distributions of cold dark matter (left) and "galaxies" (right) in an N-body simulation of an Einstein-de Sitter universe dominated by cold dark matter. The side of the box is 32.5 h^{-2} Mpc.

Matching to observations of galaxy clustering on megaparsec scales fixes the amplitude of the power spectrum and the values of h, s and ν. There are no further free parameters in the flat CDM model and naturally the next question to address is whether it can also account for observations on other scales. We first consider scales larger than $\sim$10 Mpc and then galactic scales. In what follows we adopt the model parameters that fit the clustering data: h=0.5, s=1 Mpc and ν=2.5.

3.2 Clusters, filaments and voids

To investigate the very large scale behaviour of a flat CDM model we ran a series of N-body simulations of cubic comoving regions of present size 280 and 360 Mpc in which the galaxies are assumed to be biased in the manner described above [49]. The resolution of these calculations ($\sim$1 Mpc) is insufficient to find the peaks of the smooth linear density field directly, but we can circumvent this difficulty using the analytic machinery which Bardeen et al [46] developed to study the statistics of peaks in Gaussian random fields. This formalism enables one to calculate the expected number of high peaks as a non-linear function of the density field smoothed on a large scale that we can resolve. We located "galaxies" in the simulations by assuming their abundance to be proportional to the local number density of peaks. This procedure allows us to associate a total luminosity with each particle in the simulation; we further assumed that the probability of finding a galaxy of a given luminosity associated with a particle is given by Schechter's universal luminosity function with the parameter values recommended by Felten [50]. In this way we can construct artificial volume and magnitude limited "galaxy" catalogues analogous to those used in observational studies.

To obtain the predicted properties of galaxy clusters we·identified rich clusters in the simulations using a criterion patterned after the one which Abell [51] used to find clusters in the Palomar Sky Survey. We found that the model produces the observed abundance of clusters both of richness class greater than 0 and of richness class greater than 1. This agreement is remarkable since the model has no adjustable parameters. The model clusters have masses and luminosities typical of clusters in the real Universe. The predicted mass to light (M/L) ratios inside the "Abell radius" are essentially independent of cluster radius and decrease slowly with cluster luminosity. The mean mass to light ratio of the clusters is about 20% of the value required to close the universe; thus virial analyses of these clusters would lead to the same low values of Ω inferred from real clusters. The mean M/L of the model clusters does, of course, increase as we average their properties to larger and larger radii, but the bias remains large over regions the size of the Local Supercluster. The estimates of $\Omega \sim 0.3$ obtained from standard analyses of our infall velocity towards the Virgo cluster are exactly what would be expected in a flat cold dark matter universe with a bias of this kind. The only serious discrepancy that we found between our models and exisiting observational data on galaxy clusters is the amplitude of cluster-cluster clustering. In our models the cluster autocorrelation function falls to unity at about 22 Mpc, a factor of 2.2 less than in the sample of $\sim$100 Abell clusters analysed by Bahcall and Soneira [52] and Klypin and Kopylov [53]. This difference is significant at the 2σ level [49,54] and may present a serious difficulty for the model; however systematic effects in Abell's catalogue cannot be excluded at this time. Larger and more homogeneous cluster samples are clearly very desirable in order to confirm the observational result.

We also calculated the predicted values of various velocities of interest. The rms peculiar velocity of clusters is 145 km/s, in agreement with the upper limits derived by Aaranson et al [55]; the rms relative peculiar velocity of cluster pairs closer than 56 Mpc is 511 km/s, much smaller than the velocity broadening of $\sim$2000 km/s claimed by Bahcall and Soneira [52] in their correlation study. (Both values refer to velocities in 3 dimensions). It is not clear, however, that these two observational results are consistent with one another and again a larger sample would be desirable. Another potentially serious problem which may be related to the weak cluster-cluster clustering in the model is that the low peculiar velocities that we find both for galaxies and for clusters are inconsistent with the large scale streaming motions recently reported [56,57]. However, the rms peculiar velocity of "galaxies" in our models is 490 km/s, quite comparable to our observed velocity through the microwave background.

Using our artificial volume limited catalogues we examined the distribution of voids predicted in the model. In the simulations it is relatively straightforward to compute, for example, the distribution of the size of the largest empty sphere surrounding a randomly chosen point, but no comparable data exists for the real Universe. The only available datum is the large void in Bootes, with a diameter of 124 Mpc, discovered by Kirshner et al [58,59] using redshift surveys on narrow beams arranged in a regular mesh. Following their procedure quite closely we found that the CDM model readily produces voids approaching this size. In addition simulations of deep pencil beam surveys show

striking voids and generally similar structures to those found
observationally [60,61].

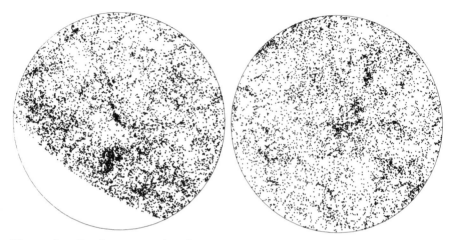

Figure 5. Equal area projections of the galaxy distribution in the
North Galactic Cap. Left: The Zwicky catalogue. Right: A magnitude
limited survey from a simulation of a flat cold dark matter universe [49]

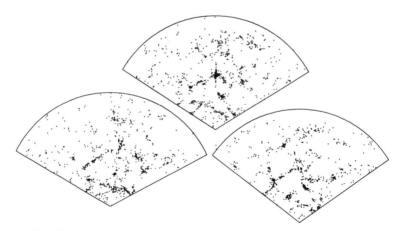

Figure 6. Wedge diagrams from an N-body simulation of a flat cold dark
matter universe. The radial coordinate is redshift distance and the
angular coordinate is right ascension. The diagrams at the bottom left
and bottom right correspond to 6° declination strips immediately below
and above the 6° strip shown at the top.

    Artificial magnitude limited catalogues with an apparent magnitude
limit $m_{zwicky} < 15.5$ show a similar morphology to the real Zwicky
catalogue when projected on the sky (fig. 5). Filaments, large low
density regions and apparent superclusters are seen in the real map as

well as in the artificial data. From this map we can extract a strip
between declinations 26.5° and 32.5° and construct a "redshift survey"
analogous to the real one recently completed by de Lapparent et al [62]
who reported the detection of "striking... surfaces of bubble-like
structures" of typical diameter 50 Mpc. An artificial wedge diagram in
redshift space of the 6° strip is shown in fig. 6, together with similar
diagrams for the strips immediately above and below it. Some of the
filaments can be traced from one strip to the next suggesting that they
are, in fact, sections through sheets. The long curving filaments often
with sharp edges have a strong resemblance to the real data. Note that
several of these filaments are of the order of 100 Mpc long and so are
quite comparable to the Perseus-Piscis chain, the largest known
filamentary structure [63,64].

3.3 Galactic halos

If it is to be viable the cold dark matter model should also be
able to account for the abundance and characteristic properties of
galaxy halos. To study this problem, we recently carried out a series of
high resolution N-body simulations [42,43,65]. We followed the
evolution of 9 small cubic regions of present size $2(1+z_i)$ Mpc, from
time $t_i = 7.2 \times 10^8$ yrs (corresponding to z=6 for $\Omega=1$) to the present day.
The resolution is determined by the softening parameter and corresponds
to ∿2kpc at the start of the calculations and ∿14 kpc at the end. Thus,
these models are able to resolve the internal regions of galaxies but
neglect the effect of large wavelength perturbations. These can be
incorporated in an approximate way by varying the mean density of the
simulations. So long as these large wavelengths remain linear, a region
of higher than average density will evolve as a closed universe, whereas
a region of lower than average density will evolve as an open
universe. Three of our calculations (ensemble EFLAT) had the mean
cosmological density, three (ensemble ECLOSED) a density slightly above
average and the remaining three (ensemble EOPEN) a density slightly
below average. The expansion factors were chosen so that all regions
have the same age; Ω varies from 0.85 to 0.41 in EOPEN and from 1.17 to
4.79 in ECLOSED. The simulations all have 32768 particles corresponding
to a mass per particle of 5.8, 3.9, and 14.3 (10^5 $M_\odot$) in EFLAT, EOPEN
and ECLOSED respectively.

Before z=∿3 the mass distributions in the simulations are quite
smooth but by z=2.5 a few tens of clumps with circular speeds exceeding
100km/s have formed near the high peaks of the initial density field.
Between z=2.5 and the present the clumps undergo considerable evolution.
Some of them merge while others remain in relative isolation during
extended periods, gradually accreting surrounding material. By z=0
almost all the clumps present at z=2.5 have been disrupted and
incorporated into larger systems. The final halos are smooth, centrally
concentrated and, in most cases, significantly triaxial. The merging
activity is at its peak prior to z=1 and only about 30% of the final
halos undergo a major merger between z=0.4 and the present. About half
the halos evolve from z=1 to their final state by accretion of material
onto their outer parts. The merger rate is somewhat larger in high
density regions where the largest clumps are formed. The formation

path of one of the largest halos in the simulations is illustrated in fig. 7; it forms by a complicated series of merger events involving 6 or 7 clumps, originally distributed on sheets. An important feature of merger events is that a significant transfer of angular momentum from the orbits of subclumps to outlying material occurs as substructure is erased. For example the most bound 20 percent of the clump in fig. 7 lost 70 percent of its angular momentum between a redshift of 1 and the present. As a result, the inner parts of merger remnants tend to be slowly rotating [65,66]. In the CDM cosmogony violent dynamical activity is expected even at fairly recent epochs. Galaxy formation is predicted to be a recent and protracted process which continues until the present.

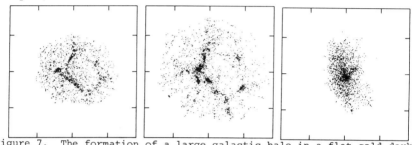

Figure 7. The formation of a large galactic halo in a flat cold dark matter universe. From left to right, projections in physical co-ordinates at redshifts of 2.5, 1.0, 0 of all the particles that end up within a region of overdensity ∿500, tick marks are at 1 Mpc intervals.

The internal structure of galactic halos can be inferred from measurements of rotation curves in the outer parts of disk galaxies [67,68]. The rotation curves of halos in the simulations, calculated as $Vc(r)=(GM(r)/r)^{\frac{1}{2}}$, are shown in fig. 8. Most are remarkably flat at radial distances beyond a few tens of kiloparsecs and resemble the rotation curves measured in the outer parts of spiral galaxies. These tend to remain flat to smaller radii than we can resolve in the calculations; however, the gravitational effects of the luminous

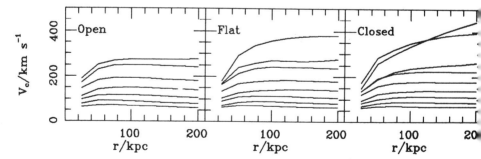

Figure 8. Average rotation curves for halos in the three ensembles at the end of the simulations. Curves were binned in logarithmic intervals of the asymptotic circular velocity, V_{max} and only halos with $V_{max}>63$ km/s are included.

material collecting onto the centre are likely to extend a flat rotation curve into the inner regions of the galaxy [69-71]. The deepest potential wells in ECLOSED have rotation curves which continue to rise steeply beyond $\sim$100 kpc and have amplitudes well above those inferred for the halos of most spiral galaxies. While some of them may correspond to the massive halos of bright ellipticals, others appear to be much too massive to be associated with any single galaxy and should perhaps be identified with the merged halos of small groups of galaxies. This cannot be decided with the present simulations since they do not include the dissipational physics required to separate baryonic material from dark matter. Small galaxy groups, however, are observed in the real Universe and our results suggest that for the cold dark matter cosmogony to be viable, the galaxies must merge on a much longer timescale than their individual halos.

From the simulations we can readily calculate the predicted number density of halos as a function of characteristic velocity V_{max} and of mean background density. The abundance of galactic halos in the real Universe can be inferred from the galaxy luminosity function and the empirical relation between luminosity and characteristic velocity (the Tully-Fisher relation). The model predictions match the observations very well, a remarkable agreement since there are no adjustable parameters in the calculations [42]. It appears that the CDM model can reproduce the observed abundance of bound structures ranging from galaxy halos to galaxy clusters. At any given V_{max}, the number density of halos is smaller in the low density models than in the high density ones; no halos with $V_{max} > 275$ km/s were formed in EOPEN, whereas a small number of halos with $V_{max} > 400$ km/s was found in ECLOSED.

3.4 Biased galaxy formation

We have seen that large halos tend to form preferentially in regions where the local mass density is high. This bias is of dynamical origin and is caused by the different rates at which non-linear gravitational clustering proceeds in regions of different density. It raises the attractive possibility that the "ad hoc" bias in the galaxy distribution invoked to reconcile a flat universe with observations may actually arise naturally in the CDM model. To test this possibility we compare the abundance and clustering properties of galactic size halos in our simulations with those predicted by the high-peak model. Using the formalism of Bardeen et al. [46] we can calculate the excess of 2.5σ peaks in ECLOSED relative to EOPEN expected in this model and compare it to the number of halos actually found. This is about 1/2 the expected number, but it represents only a lower limit to the expected bias in the population of bright galaxies since, as discussed above, the largest halos must correspond to groups. For example, groups of $\sim$5 galaxies in halos with $V_{max} > 400$ km/s would account for all the expected bias.

To examine the spatial correlations of halos we ran a simulation in a bigger box (32 Mpc) with 110592 particles and slightly poorer resolution than the ensembles. There is again an ambiguity in the comparison with observations because underestimating the number of galaxies in merged halos will depress their correlations. This effect can be crudely incorporated in the data by smoothing the galaxy counts within a velocity difference of 300 km/s and projected separation of $\sim$ 1.0 Mpc in the CfA catalogue. These group selection scales correspond roughly to the size of the largest halos in the simulations. By replacing galaxies within such groups by the group centres we find that the correlation length of the data drops to the value estimated for halos with V_{max} >200 km/s in the simulations. Thus, provided large halos contain small galaxy groupings, the natural bias seems to be enough to account for the requirements of an Ω=1 CDM universe. In this case, less luminous galaxies would be expected to be less clustered than brighter galaxies.

4. CONCLUSIONS

The standard hot big bang cosmology has been complemented in recent years by new ideas from particle physics concerning the very early Universe and the possible identity of the missing mass. The inflationary model has reinforced the old prejudice that the Universe ought to be flat and has provided a mechanism to generate primordial density fluctuations. The possibility that the "missing mass" may consist of weakly interacting elementary particles has led to specific predictions for the galaxy distribution that can be tested against observations. Linear analyses of the growth of density perturbations provide the input to N-body simulations which can accurately follow the late non-linear phases of evolution. These and other related studies have shown that light stable neutrinos with a mass of about 30 ev are unlikely candidates for the missing mass. On the other hand, cold collisionless relics (such as supersymmetric particles, Gev neutrinos or axions) appear quite promising.

An open model in which bright galaxies trace the distribution of cold particles is compatible with observations of galaxy clustering on small scales but appears to be inconsistent with the observed isotropy of the microwave background. A flat model can be reconciled with small-scale dynamical determinations of Ω if galaxies form preferentially in regions of high mass density. A bias in this sense occurs naturally in a cold dark matter universe because the binding energies of galactic size objects are significantly affected by large-scale fluctuations. N-body simulations show that galactic halos are indeed more strongly clustered than the overall mass distribution and that this effect may be enough to account for the requirements of an Ω =1 universe; however the present results are not conclusive because they do not include the dissipative effects required to resolve individual galaxies within merged halos. In the mean, the flat cold dark matter model produces the observed abundance of galactic halos with flat rotation curves and amplitudes similar to those inferred for the massive halos of spiral galaxies. Galaxy formation is predicted to be a recent but protracted process continuing until the present day. Mergers play an important role and may explain the relative abundance and slow rotation of bright elliptical galaxies. Using a simple prescription for the bias in the galaxy distribution we have shown that the flat cold dark matter model is consistent with measures of galaxy clustering on intermediate scales, such as the 2- and 3- point correlation functions and the peculiar velocities of galaxy pairs. On larger scales, the model is consistent with the luminosity function, abundance, mass and mass to

light ratio of Abell clusters, with our inferred infall velocity towards
Virgo, with voids as large as the one in Bootes, with filaments similar
to the Perseus-Piscis chain, with the "bubbles" and general morphology
seen in recent redshift surveys, and with our observed motion through
the microwave background. Significant discrepancies are found only on
the largest scales where the model may not produce the level of
superclustering implied by recent determinations of cluster-cluster
clustering and large-scale streaming motions. Thus the cold dark matter
cosmology reproduces the abundance and gross properties of observed
virialized structures spanning the mass range between galaxy halos and
rich galaxy clusters. This agreement with observations is remarkable,
not least because it involves no adjustable parameters other than the
scaling factors of the primordial fluctuation spectrum which are fixed
by the match to observations of galaxy clustering on intermediate
scales.

Ultimately the question of what is the missing mass can only be
conclusively resolved by direct detection. The success of the flat cold
dark matter cosmology should provide strong encouragement to current
efforts aimed at detecting these particles.

Acknowledgements.

Most of the material presented here is joint work carried out with
George Efstathiou, Marc Davis and Simon White whom I thank for
permission to present some of our new results prior to joint
publication. Part of this work was supported by NATO travel grant
689/84.

References

1. Guth, A., Phys. Rev. D23, 347 (1981).
2. Starobinskii, A., Phys. Lett. 117B, 175 (1982).
3. Bardeen, J.M., Steinhardt, P.J. and Turner, M.S.,Phys. Rev. D28,
   679 (1983).
4. Bond, J.R., Efstathiou, G. and Silk, J., Phys. Rev. Lett. 45, 1980
   (1980).
5. Guyot, M. and Zel'dovich, Ya.B., Astron. Astrophys. 9, 227 (1970).
6. Meszaros, P., Astron. Astrophys. 37, 225 (1974).
7. Bond, J.R., Centrella, J., Szalay, A.S. and Wilson, J.R., In
   Formation and evolution of galaxies and large structures in the
   Universe eds J. Audouze and J. Tran Thanh Van (Dordrecht: Reidel),
   p.87 (1983).
8. Primack, J.R., Lectures presented at the International School of
   Physics "Enrico Fermi" (Varenna, Italy) (1984).
9. Peebles P.J.E., Astrophys. J. 258, 415 (1982).
10. Bond, J.R. and Szalay, A.S., Astrophys. J. 274, 443 (1983).
11. Witten, E., Phys. Rev. D30, 272 (1984).
12. Carr, B.J., Comments on Astrophys. 7, 161 (1978).
13. Pagels, H.R. and Primack, J.R., Phys. Rev. Lett. 48, 223 (1982).
14. Bond, J.R., Szalay, A.S. and Turner, M.S., Phys. Rev. Lett. 48,
    1636 (1982).

78

15. Efstathiou, G., Davis, M. Frenk, C.S. and White, S.D.M., Astrophys. J. Suppl. 57, 241 (1985).
16. Frenk, C.S., Phil. Trans. Roy. Soc., in press (1986).
17. White, S.D.M., In Inner Space/Outer Space, ed. E.W. Kolb, M.S. Turner, D. Lindley, K. Olive, D. Seckel (Univ. Chicago Press) p.228 (1986).
18. Lyubimov, V.A., Novikov, E.G., Nozik, V.Z., Tretyakov, E.F. and Kozik, V. F., Phys. Lett. 94B, 266 (1980).
19. Reines, F., Sobel, H. and Pasierb, E., Phys. Rev. Lett. 45, 1307 (1980).
20. Frenk, C.S., White, S.D.M. and Davis, M., Astrophys. J. 271, 417 (1983).
21. White, S.D.M., Frenk, C.S. and Davis, M., Astrophys. J. 274, L1 (1983).
22. White, S.D.M., Davis, M. and Frenk, C.S., Mon. Not. R. astr. Soc. 209, 27p (1984).
23. Klypin, A.A. and Shandarin, S. F., Mon. Not. R. astr. Soc. 204, 891 (1983).
24. Centrella, J. and Melott, A.L., Nature 305, 196 (1983).
25. Cowie, L. Paper delivered at the CITA conference on Galaxy Formation. (Toronto, Canada) (1985).
26. Davis, M., Huchra, J., Latham, D.W. and Tonry, J., Astrophys. J. 253, 423 (1982).
27. Davis, M., and Huchra, J., Astrophys. J. 254, 425 (1982).
28. Doroshkevich, A.G. and Khlopov, M. Yu., Soviet Astron. Lett. 9, 171 (1983).
29. Hut, P. and White, S.D.M., Nature 310, 637 (1984).
30. Turner, M.S., Steigman, G. and Krauss, L.M., Phys. Rev. Lett. 52, 2090 (1984).
31. Schramm, D.N., In Proc. of the Bielefield Conference on Phase Transitions in the Early Universe, in press (1986).
32. Efstathiou, G., Mon. Not. R. astr. Soc. 213, 29p (1985).
33. Kaiser, N., Astrophys. J. 273, L17 (1983).
34. Faber, S.M. and Lin, D.N.C., Astrophys. J. Lett. 266, L17 (1983).
35. Blumenthal, G.R., Faber, S.M., Primack, J.R. and Rees, M.J., Nature 311, 517 (1984).
36. Davis, M., Efstathiou, G., Frenk, C.S. and White, S.D.M., Astrophys. J. 292, 371 (1985).
37. Bond, J.R. and Efstathiou, G., Astrophys. J. 285, L45 (1984).
38. Vittorio, N. and Silk, J., Astrophys. J. 285, L39 (1984).
39. Uson, J.M. and Wilkinson, D.T., Astrophys. J. 277, L1 (1984).
40. Rees, M.J., Mon. Not. R. astr. Soc. 213, 75p (1985).
41. Silk, J., Astrophys. J. 297, 1 (1985).
42. Frenk, C.S., In Nearly Normal Galaxies: from the Planck Time to the Present, ed. S. Faber (Springer-Verlag) in press (1986).
43. Frenk, C.S., White, S.D.M., Davis, M. and Efstathiou, G., in preparation (1986).
44. Schaeffer, R. and Silk, J., Astrophys. J. 292, 319 (1985).
45. Bardeen, J.M., In Inner Space/Outer Space, ed. E.W. Kolb, M.S. Turner, D. Lindley, K. Olive, D. Seckel (Univ. Chicago Press) p.212 (1986).
46. Bardeen, J.M., Bond, J.R., Kaiser, N. and Szalay, A.S., Astrophys. J. 304, 15 (1986).

47. Kaiser, N., In Inner Space/Outer Space, ed. E.W. Kolb, M.S. Turner, D. Lindley, K. Olive, D. Seckel (Univ. Chicago Press) p.258 (1986).
48. Kaiser, N., Astrophys. J. 284, L9 (1984).
49. White, S.D.M., Frenk, C.S., Davis, M. and Efstathiou, G., Astrophys. J. in press (1986).
50. Felten, J.E., Comm. Astrophys. Sp. Sci. 11, 53 (1985).
51. Abell, G.O., Astrophys. J. Suppl. 3, 211 (1958).
52. Bahcall, N. and Soneira, R.M., Astrophys. J. 270, 20 (1983).
53. Klypin, A.A. and Kopylov, A.I., Sov. Astron. Lett. 9(1), 41 (1983).
54. Ling, N., Frenk, C.S. and Barrow, J., Mon. Not. Roy. astr. Soc., in press (1986).
55. Aaronson, M., Bothun, G., Mould, J., Huchra, J.P., Schommer, R.A. and Cornell, M.E., Astrophys. J. 302, 306 (1986).
56. Collins, A., Joseph, R.D. and Robertson, N.A., Nature, in press (1986).
57. Burstein, D., Davies, R.L., Dressler, A., Faber, S.M., Lynden-Bell, D., Terlevich, R. and Wegner, G., preprint (1986).
58. Kirshner, R.P., Oemler, A. Schechter, P.L. and Shectman, S. A., Astrophys. J. 248, L57 (1981).
59. Kirshner, R.P., Oemler, A. Schechter, P.L. and Shectman, S. A., In IAU Symposium 104, Early evolution of the Universe and its present structure eds G.O. Abell and G. Chincarini (Dordrecht: Reidel) p.197 (1983).
60. Broadhurst, T., Ellis, R.S. and Shanks, T., in preparation (1986).
61. Koo, D., Kron, R. and Szalay, A.S., in preparation (1986).
62. de Lapparent, V., Geller, M.J. and Huchra, J.P., Astrophys. J. (Letters) 302, L1 (1986).
63. Gregory, S.A. and Thompson, L.A., Astrophys. J. 222, 784 (1978).
64. Chincarini, G.L., Giovanelli, R. and Haynes, M.P., Astron. Astrophys. 121, 5 (1983).
65. Frenk, C.S., White, S.D.M., Efstathiou, G. and Davis, M., Nature 317, 595 (1985).
66. Barnes, J. and Efstathiou, G., preprint, Univ of Cambridge (1986).
67. Rubin, V.C., Burstein, D., Ford, W.K. and Thonnard, N., Astrophys.J. 289, 81 (1985).
68. van Albada, T.S., Bahcall, J.N., Begeman, K. and Sanscisi, R. Astrophys. J. 295, 305 (1985).
69. Gunn, J.E., In Dark Matter in the Universe, ed G. Knapp (Dordrecht: Reidel) in press (1986).
70. Blumenthal, G.R., Faber, S.M., Flores, R.A. and Primack, J., Astrophys. J. 301, 27 (1986).
71. Barnes, J. In Nearly Normal Galaxies: from the Planck Time to the Present, ed. S. Faber (Springer-Verlag) in press (1986).

Bounds on Galactic Cold Dark Matter Particle Candidates and Solar Axions from a Ge-spectrometer

presented and written by
Graciela Gelmini[*]

Lyman Laboratory of Physics, Harvard University
Cambridge, MA 02138
USA

based on work done in collaboration with
S. P. Ahlen,[1] F. T. Avignone III,[2] R. L. Brodzinski,[3] S.
Dimopoulos,[4] A. K. Drukier,[5] B. W. Lynn,[4] D.N. Spergel[6]
and G.D. Starkman[4]

ABSTRACT

The ultralow background Ge spectrometer developed by the USC/PNL group is used as a detector of cold dark matter candidates from the halo of our galaxy and of solar axions (and other light bosons), yielding interesting bounds. Some of them are: heavy standard Dirac neutrinos with mass $20 \ GeV \leq m \leq 1 \ TeV$ are excluded as main components of the halo of our galaxy; Dine-Fischler-Srednicki axion models with $F/2x_e' \leq 0.5{\times}10^7 \ GeV$ are excluded.

1. Introduction

It is by now a quite well established fact that most of the matter of the universe is non-luminous, even if the total amount and the nature of this "dark" matter (D.M.) are still fascinating open questions.[1] The rotation curves of spiral galaxies indicate that most of their mass is in a dark approximately spherical halo. Even if it is not excluded that the D.M. consists of nucleonic matter, in the form of faint stars for example, there are several arguments against it. An attractive possibility is that the D.M. is neutral elementary particles. Some particle

---

[*] Address from October 1986: The Enrico Fermi Institute, University of Chicago, Chicago, IL 60637; on leave of absence from Department of Physics, University of Rome II, Via Orazio Raimondo, Rome, Italy, 00173. Work partly supported by the U.S. Department of Energy, Grant No. DE AC02 82ER-40073, and NSF Grant PHY-82-15249.
[1] Dept. of Physics, Boston University, Boston, MA 02215.
[2] Dept. of Physics, Univ. of South Carolina, Columbia, SC 29208.
[3] Pacific Northwest Laboratory, Richland, WA 99352.
[4] Dept. of Physics, Stanford University, Stanford, CA 94305.
[5] Applied Research Corp., 8201 Corporate Dr., Landover, MD 20785.
[6] Institute for Advanced Studies, Princeton, NJ 08544.

candidates are already considered standard, such as a light or a heavy massive neutrino, the axion, the lightest supersymmetric particle...[1], and many more could be proposed. In this case we may be able to detect the D.M. particles which constitute the halo of our own galaxy, depending on their masses and their interactions with the particles we already know. This is an impossible task for light neutrinos at present,[2] but it may be possible for "cold" D.M. Halo axions could convert to photons,[2] and heavy enought particles may be detected indirectly through their annihilation products (see the contributions of M. Srednicki and K. Olive to this workshop). Heavy enough particles may also be detected directly, through the energy they would deposit in collisions within detectors. Several supercooled bolometric detectors for this purpose are under study.[2] (See Ref. 2 and the contribution of L. Stodolsky to this workshop). This talk refers to the first experiment based on direct detection, performed with an already existing (non-supercooled) germanium detector[3], with which also bounds on solar axions were obtained.[4]

2. Direct Detection of Galactic Cold D.M. Particles

The idea of direct detection is that an incident halo particle could collide with a nucleus and transfer an energy that could be measured.[5,6,7] Typical velocities of D.M. particles in the halo should be $v \simeq 10^{-3}c$. Since the threshold energy expected, in any detection method considered up to now, is around 1 keV, direct searches can test particles of masses $m \geq 1$ GeV. The maximum energy transfer in a collision of a projectile particle with velocity v with a target of mass M at rest is $T^{\max} = 2Mm^2/(m + M)^2(v/c)^2$. The non-relativistic cross-section for this process is $\sigma = \dfrac{m^2M^2}{\pi(m + M)^2}\,|A|^2$ where A is a reduced amplitude which depends on the dynamics of the collision. Due to the value of m, both $T^{\max}$ and σ, point towards nuclei as good targets, as opposed to electrons. If D.M. particles could be detected and nuclear recoil spectra of several materials could be measured then m and v of the particles could be estimated from the recoil energy cut-off $T^{\max}$. The type of interactions could be inferred from the different shapes of the spectra: for coherent interactions A should scale with the atomic number of the target (or some number in relation to it, like the number of neutrons for weak interactions); for spin-dependent interactions A depends on the total spin of the target. The proof that the particles detected belong to the halo of our galaxy would be the annual modulation of the signal.[7] Due to the motion of the earth around the sun, and the fact that the sun is not at rest with respect to the signal, it should be maximum in June and minimum six months later. A halo consisting of "cold" D.M. can not be rotating with the disc of the galaxy or otherwise it would be flattened like the disc. (An upper bound is provided by the rotation of the bulge, which is slower than the disc, even if it is probably that the halo moves much more slowly or not at all). The sun moves with the disc. While the disc is supported against radial collapse by its angular momentum, the dark halo is supported by random velocity. When the matter that makes the halo collapsed to form our galaxy, the velocities of its constituents were randomized in a process called "violent relaxation". Since the D.M. particles are practically collisionless they retain the original velocity distribution. This mechanism does not insure a random mixing at small scales. But at small scales the phase mixing is due to the fact that the D.M. particles have already performed several hundred oscillations in the galactic potential starting from slightly different initial conditions. The local total velocity dispersion $v_{r.m.s.}$ can be inferred from the observed local circular speed; it it is of order $10^{-3}c$, similar to the velocity of the sun with respect to the halo $v_{\odot-h}$. Thus, most particles have velocities between zero and $2v_{r.m.s.}$ with respect to the earth. Due to the existence of an energy threshold, that translated into a velocity threshold $v_{th} > 0$, only particles with velocities $v > v_{th}$ can be detected. That is why, when the relatively small velocity of the earth around the sun, of order $10^{-4}c$, adds up to $v_{\odot-h}$ in June, many particles become visible, and six months later many become undetectable. The amplitude of the signal modulation depends on the detector velocity threshold v_{th}.[7]

The range of masses of few GeV is particularly interesting to test. Particles with weak (or close to weak) interactions whose cosmological abundance ρ is determined by their annihilation in the early universe, should have masses in that range to account for the halos of galaxies or to be the dominant form of matter in the universe, $\rho \simeq (0.1$ to $1.0)\rho_c$. For example, heavy Dirac neutrinos should have a mass $m \simeq 5$ to $10\ GeV$,[8] Majorana neutrinos a mass $m \simeq 10$ to $20\ GeV$,[8] sneutrinos a mass $m \leq 10\ GeV$ [9] and photinos a mass $m \simeq 2\text{--}5\ GeV$ [10] to account for $\rho \simeq 0.1\ \rho_c$ (where ρ_c is the critical density, which just closes the universe). Dark matter bounded to the disc of our galaxy would have velocities of order $10^{-4}c$ with respect to the earth, too low to be detected with particles in the GeV range. Only masses larger than $100\ GeV$ could be tested. However, for very heavy particles $m >> M$, the recoil energy does not increase with m. Thus $T^{max} \simeq 1\ keV\ (M/100\ GeV)$ depends only on the mass M of the nucleus and is never much larger than $1\ keV$.

A recoiling nucleus in a crystal gives part of its kinetic energy to the lattice to produce a phonon wave (balistic phonons) which, after many scatterings within the restricted volume of the target, thermalize originating an increase in temperature. Experiments in which either balistic phonons or small temperature variations are detected are under study. They require cryogenic methods, with temperatures of order $10 - 100\ m$ Kelvin. The heat capacity of materials, in particular pure crystals or superconductors decrease rapidly with the temperature. Thus at lower temperatures smaller energy recoils are necessary to produce a given temperature increase. (See ref. 2 and the contribution of L. Stodolsky to this workshop.)

Another effect of a recoiling nucleus in a crystal is the production of ionization within the material. Part of the nuclear recoil energy goes into the excitation of electrons. This mechanism allows the use of an existing ultra low background germanium diode detector to search for the galactic D.M. This type of detector was developed to look for the double β decay of ^{76}Ge for what an excellent background rejection and high sensitivity are needed. These properties, together with a low energy threshold, are essential in the search for D.M. This detector counts the number of electrons which jump into the conduction band, i.e., the number of "shallow" electron-hole, $e-h$, pairs as a result of energy deposited either in an electron or in a nucleus of the Ge crystal. The low band gap ($0.69\ eV$ at 77^0K) and high efficiency for converting electronic energy loss to electron-hole ($e-h$) pairs ($2.96\ eV$ per $e-h$ pair at 77^0K), make Ge detectors probably the best existing particle detectors utilizing low velocity recoiling nuclei.

3. Limits on Halo Cold D.M. Candidates from the PNL/USC Ge Spectrometer

The PNL/USC group has developed a $135\ cm^3$ prototype intrinsic Ge detector having a background reduced by about three orders of magnitude over conventional low background gamma ray spectrometer, the lowest background ever achieved.[11] The detector is at a depth of 4000 meters of water-equivalent in the Homestake goldmine, in order to eliminate the cosmic ray induced background. The Ge crystals themselves are free from primordial or man-made radioactivity and the materials used in the construction of the detector and shields were carefully studied and selected. The data on which the present analysis is based were taken with the detector surrounded by high-purity copper and 11 tons of lead to eliminate the radiative background from the rock. Recently the shield has been upgraded by the use of 448 year-old lead (from a sunken Spanish galleon) replacing the copper inner shield which had cosmogenic radioactivty. When the seach for dark matter was initiated in late 1985 the energy threshold was reduced to an incident electron energy of $4\ keV$. This corresponds to an initial nuclear recoil energy of $15\ keV$. Below this threshold there was a strong increase of microphonic noise engendered by mining operation. Hardware and software have recently been developed to reduce this noise and bring the $4\ keV$ threshold to $1\ keV$. The number of $(e-h)$ pairs produced when some energy T is deposited on an electron is a well known property of the Ge (approximately $T/3\ eV$). This number is larger than the one produced when the same energy is deposited on a nucleus. Let us call R.E.F., relative efficiency factor, the ratio of these two numbers. This is an energy dependent quantity. It is ~1 only for $T \gg 1\ MeV$. At smaller energies it is larger than 1 and could become very large at some energy below the present threshold. If it is so, the energy at which that happens will be the final threshold achievable to detect nuclear recoils (since the number of $e-h$ pairs produced becomes negligible). The

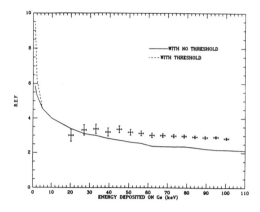

FIG. 1. Energy dependence of relative
efficiency factor (REF). The data points
are from Ref. (11).

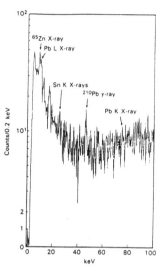

FIG. 2 (above). Ten weeks of
data from the Ge spectrometer
(0.2 keV per channel). The
identified peaks resulted from
the decay products of radiac-
tivity in one solder point
(now removed).

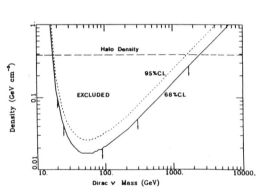

FIG. 3 (left). Maximum halo
density of standard Dirac
neutrinos consistent with the
observed count rate at the
68% and 95% confidence levels.

FIG. 4 (right). Excluded regions in mass/cross section
space $(g/g_{weak}=\sqrt{\sigma/\sigma_{weak}})$ at
the 68% C.L. for particles
constituting the halo. σ_{weak} is
the cross section of Dirac and
Majorana neutrinos for spin
independent and spin dependent
interactions, respectively.

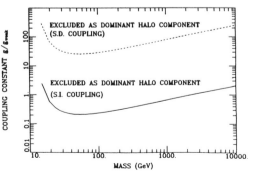

R.E.F. for Ge detectors, shown in Fig. 1 as function of the nuclear recoil, was calculated (by Steve Ahlen) by evaluating the fraction of primary Ge recoil energy lost in electronic collisions. The calculation included the electronic loss of the second and higher order generations of recoiling nuclei (the ones perturbed by the primary nucleus). The theoretical values are in good agreement with the experimental data from Chasman et al.,[12] also shown in Fig. 1, who measured $e-h$ yields of neutrons induced recoil nuclei within a Ge-detector. Two different assumptions on the behavior of the electronic energy loss as function of the velocity of the recoiling Ge nucleus give different results at low energies: the solid curve assumes a linear dependence, the dashed one a kinematic threshold at 0.27 keV of nuclear recoil energy (at which a Ge nucleus transfers a maximum energy of 0.7 keV to an electron in a direct collision).

The count rates of ten weeks of low energy data from the Ge spectrometer are shown in Fig. 2, as functions of energy deposited on electrons (which should be multiplied by the appropriate R.E.F. to obtain the coresponding nuclear recoil energy). To know what these data mean for hypothetical dark matter particles, the observed count rate has to be compared with the predicted rate if the halo consisted of those dark matter particles. This last rate R_P depends on the scattering cross-section σ of the D.M. particles considered on a Ge nucleus, on their local density (in number) n_x and on the velocity distribution $f(\vec{v})$ of the particles:

$$R_P(T) = n_x \Delta T \int \frac{d\sigma}{dT} (v,T) f(\vec{v}) v d^3 v \ .$$

Here T is the recoil energy and ΔT is the interval, centered in T, of recoil energies detected in a given channel.

The model of halo used assumes a local halo density of $0.01 \, M_\odot$ $/pc^3 = 0.38 \, GeV/cm^3$, and a isotropic gaussian distribution $f(\vec{v})$ with an $r.m.s.$ of $250 \, km/sec$ and a maximum of $550 \, km/sec$, both conservative velocity values. The halo is assumed to rotate slowly like the galactic spheroid, with a local velocity of $80 \, km/sec$. This is again a conservative assumption which reduces relative velocities, since the sun moves in the same sense around the galaxy at $250 \, km/sec$. The integral was evaluated over all velocity phase space and a bound on n_x was obtained for every T. The most restrictive T-dependent bound was taken as the final bound on n_x. The best bounds come from values of T near threshold because of the rapid decrease of the predicted rate with increasing T. (Notice that in the non relativistic limit $(\frac{d\sigma}{dT})^{NR} = \frac{\sigma^{NR}}{T^{max}}$) where σ_{NR} does not depend on T, and for a given T one must sum over all $T_{max} > T$). To fix ideas, stable standard Dirac and Majorana neutrinos were chosen to present the bounds, as examples of particles with coherent and spin-dependent interactions, respectively. The elastic cross sections of Dirac neutrinos of mass m and incident energy $E = m(c^2 + \dfrac{v^2}{2})$ on a nucleus of mass m_N with Z protons and N nucleus, at rest, through Z^0 exchange is:

$$\frac{d\sigma}{dT} = \frac{G_F^2 m_N c^2}{8\pi \, v^2} \, [Z(1 - 4\sin^2\theta_W) - N]^2 \, [1 + (1 - \frac{T}{E})^2 - \frac{(M_N T + m^2)}{E^2}]$$

$$\cdot \exp(-m_N \, TR^2/3)$$

where G_F is the weak coupling constant and θ_W is the weak mixing angle. The experimental factor is a nuclear form factor derived from the assumption of a gaussian density of nucleons with a nucleus of radius $R \simeq 1.2 \, A^{1/3}$ fermi, where $A = Z + N$. A better model of the nucleonic density, such as a Saxon-Woods density, produce a less steeply decreasing form factor. Thus, the use of the exponential is conservative. Only when $\sqrt{2m_N T} \cdot R \ll 2$, that is $T \ll 11 \, keV$ for Ge, the difference in phase of the scattered wave function at the location of different nucleons can be neglected and the scattered particle interacts coherently with a point-like nucleus. Fig. 3 shows the maximum halo density of heavy standard Dirac neutrinos that is consistent with the observed count rate at the 68% (solid line) and 95% (dashed line) confidence level. Standard Dirac neutrinos of mass $20 \, GeV < m < 1 \, TeV$ are excluded as main components of the halo. This figure can be used for other vectorial spin-independent interactions by scaling the vertical axis by the ratio of cross-sections. A particle with a cross section 2 times larger should have a density 2 times lower to obtain the same interaction rate, for example. This is what happens in the case of sneutrinos.

Even if the bounds found in a terrestrial experiment only test the local density of D.M. this one is considered representative of the global D.M. density in the halo. We have argued before that even at small scales the velocities of the D.M. particles should be randomized, if not by violent relaxation, by phase mixing after many oscillations in the galactic potential.

Our main results are shown in Fig. 4, where the range of mass and cross-sections for particles with spin independent and spin-dependent interactions, scaling with respect to the weak interactions of standard Dirac and Majorana neutrinos as $(g/g_{weak}) \equiv \sqrt{\sigma/\sigma_{weak}}$. The bounds on spin-dependent interacting particles are not as good as for spin-independent ones because most of the Ge isotopes have zero spin. Only ^{73}Ge, with a natural abundance of 7.8%, has a nonzero spin $S = 9/2$. The bounds apply to a maximum cross-section of $\sigma \approx 10^{-28} cm^2$, for which the dark matter particles would be stopped in the earth's crust before arriving at the detector. For large masses the predicted rate decreases as m^{-1}, consequently the bounds become less stringent.

The identified low energy peaks in the present data, Fig. 2, result from decay products of ^{210}Bi in a solder point near the detector which has now been removed. The shield has been recently upgraded. Thus, with several months of counting, the background could be reduced by another factor of 10. Through reduction of the microphonic noise engendered by mining operations it will be possible to lower the threshold to $1\ keV$ of electronic initial energy. This would mean reaching at most Dirac neutinos of mass $m > 8\ GeV$.

Bounds on galactic abundances can be translated into bounds on cosmological abundances if it is assumed that the ratio of the total mass in non-baryonic matter to the total mass in baryonic matter in our galaxy $F_{Gal} = M_\nu/M_{baryon}$ is the same as in the whole universe. This may be a good assumption for cold D.M. Fig. 5 shows the upper limits on F_{Gal} under two assumptions: 1) the halo is baryonic, $M_{baryon} = M_{halo}$, and 2) M_{baryon} includes only the observed luminous baryons (stars, gas, dust, etc.)

The cosmological ratio of stable Dirac neutrinos to baryons, $F_{cosmological}$, was calculated[3] using an analytical solution to the Boltzmann equation to find the cosmological relic density $\rho^\nu_{cosmol.}$.[13] The largest value for $\rho^{baryon}_{cosmol.}$ consistent with bounds from nucleosynthesis[14], $\Omega_b h^2 < 0.035$, was used. (Here $h = .5$ to 1, $\rho^{baryon}_{cosmol.} = \Omega_b \rho_c$ and $\rho_c = 2 \times 10^{-29} \frac{g}{cm^3} \times h^2$) thus $F_{cosmol.} = \rho^\nu_{cosmol.}/(2 \times 10^{-29} \frac{g}{cm^3} \times 0.035)$. Fig. 5 shows that if $\rho_{baryon} = \rho_{halo}$ the existence of these particles is excluded for masses larger than $20\ GeV$ except in a narrow mass range near the Z^0 resonance at $m_\nu = m_{Z^0}/2$. This resonance enhances the neutrino annihilation cross-section and lowers the relic cosmological abundance. Under the same assumptions, in the case of sneutrinos a region in the mass of photino/mass of sneutrino space is excluded, as shown in Fig. 6. Sneutrinos annihilate in the early universe mainly through a diagram proportional to the Majorana mass of the zino, which is approximately the mass of the photino, $m_{photino}$. In considering the sneutrino a candidate for the dark matter we are assuming it is the lightest supersymmetric particle, thus $m_{sneutrino} < m_{photino}$. Therefore only the region above the diagonal line in Fig. 6 is relevant. While the annihilation into two Z^0 bosons is not possible, i.e., $m_{sneutrino} < m_{Z^0}$ the above mentioned diagram dominates. To obtain a density $\rho_{sneut.} = \Omega_{sneut.} \rho_c$ it is necessary that $m_{photino} \simeq 3.5\ GeV/\Omega_{sneut} h^2$. Using that $\Omega_{sneut} h^2 < 0.035 \times F_{Gal}$ (for sneut.) lower bounds for $m_{photino}$ are obtained (Fig. 6). The excluded region goes between the curves and the diagonal; only the case $\rho_{baryons} = \rho_{halo}$ gives some bounds.

4. Limits on Solar Axions (and other Light Bosons) from the USC/PNL Ge-spectrometer

Another use of this ultralow Ge detector in the new range of energies explored since the search for D.M. started, is to test the couplings to electrons of light bosons emitted from the sun.[4] The expected spectrum of any of these particles has a maximum at an energy close to the sun temperature, $1\ keV$, while the present threshold of the detector is $4\ keV$. Interesting

88

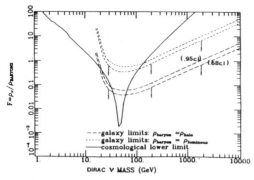

DIRAC ν MASS (GeV)

FIG. 5 (left). Maximum ratio of total Dirac neutrino mass to total baryonic mass in our galaxy if (1) all the halo is baryonic and if (2) only the visible matter is baryonic. Assuming that the ratio is the same at a cosmological scale, it can be compared with the maximum cosmological ratio consistent with the big bang nucleosynthesis (solid line).

FIG. 6 (right). Excluded region in mass of sneutrinos/mass of photino space, under the assumption that the ratio of total mass in sneutrinos over the total mass in baryons in our galaxy is the same in the whole universe. Same assumptions as in Fig. 5.

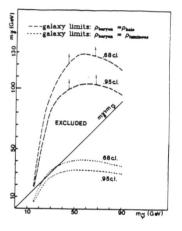

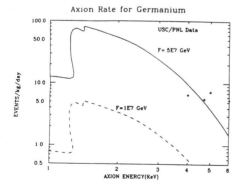

Axion Rate for Germanium

FIG. 7 (left). Solar axion events per kg per day for Ge for $F/2x_e' = 0.5 \times 10^7$ GeV (solid) and $F/2x_e' = 10^7$ GeV (dashes). The crosses are from PNL/USC data.

laboratory bounds on the coupling to electrons of pseudoscalars such as axions,[15] familons,[16] majorons[17,18] were obtained.[4] Their coupling to electrons e

$$L = g \ m_e \bar{e} \ i \gamma_s e \ \phi$$

is proportional to the electron mass m_e. When ϕ is the Dine-Fischer-Srednicki "invisible" axion, $g = 2x_e'/F$, where x_e' is a constant of order unity and F is the scale of the spontaneous breaking of the Peccei-Quinn symmetry, and the axion mass is $m_{axion} \simeq 7.2 \ eV \left[\dfrac{10^7 \ GeV}{F} \right]$. Familons[16] and singlet majorons[17] (coupled mainly to heavy right-handed neutrinos) have couplings similar to that of axions where $(2x_e')$ is replaced by a model dependent coupling constant and F is the large global horizontal and global lepton number symmetry breaking scale, respectively. Triplet-majorons[18] (coupled mainly to light left-handed neutrinos) appear if the lepton number is a global symmetry spontaneously broken at a scale v_T, the vacuum expectation value of a triplet Higgs field, small with respect to the electroweak scale. When ϕ is a triplet-majoron we have $g = 2\sqrt{2} \ G_F \ v_T$, where G_F is the weak coupling constant.

The expected rate of interactions was obtained[4] by calculating the solar axion flux, using a solar model[19,20,21] with a temperature of 1 keV, and multiplying it by the "axiolectric" cross-section on Ge atoms. Only the bremsstrahlung emission process in the sun was used, even in the case of axions[21] (the Primakoff process is suppressed due to Debye-Hückel screening in the solar plasma). The "axiolectric" effect[21] is an atomic enhancement of the interaction of scalar bosons with electrons, similar to the photoelectric effect (where the photon is replaced by a spin zero boson whose mass is zero or negligible). The absorption of a boson, forbidden by energy-momentum conservation for a free electron, becomes allowed for a bounded electron when the uncertainty in the linear momentum of the electron is larger than the momentum transferred in the absorptive process. Then, a boson will be absorbed by a bound electron which is then ejected from the atom. In the case of solar bosons, whose energies are comparable to atomic energies, around 1 keV, the enhancement is large. In the diapole approximation, for axion energies $\omega \ll m_e$ in units of $\hbar = c = 1$ we have

$$\sigma_{axiolectric} = \frac{g^2}{4\pi} \alpha_{em} \ (\frac{\omega}{2m_e})^2 \ \sigma_{photoelectric}$$

where $\alpha_{em} \simeq (127)^{-1}$.

In Fig. 7 the expected number of events per kg per day for germanium are plotted against the incoming axion energy (i.e., energy transferred to an electron) for $g^{-1} = 0.5 \times 10^7 \ GeV$ (solid line) and $g^{-1} = 1 \times 10^7 \ GeV$ (dashed line). The major contribution comes from 1 $keV < \omega < 10 \ keV$ since both the solar axion flux and the axiolectric cross-section peak in this region. The first experimental points near threshold , for $\omega \geq 4\ keV$, shown in Fig. 2, are plotted as crosses in Fig. 7. The statistical error on these data is estimated at $\pm\ 25\%$. From this the experimental bound

$$g^{-1} = \frac{F}{2x_e}, \geq 0.5{\times}10^7\ GeV$$

was deduced. This means $v_T \leq 6.9\ MeV$. Here there is a problem of conceptual self-consistency which will disappear if the bound deduced improves by a factor larger than two. In the calculations a model of the sun was used in which the dynamics was dominated by QED, weak and nuclear processes. This would not be the case if the solar axion (or other bosons) luminosity exceeds the photon luminosity. The condition that ensures this does not occur (and therefore that the sun does not burn too quickly and thus is older than $\sim 4.5{\times}10^9$ years, the age of the oldest meteorites[19,20,21]) is

$$g^{-1} \geq 1.08{\times}10^7\ GeV\quad .$$

This bound differs from that in Refs. 19 and 20, because only solar bremsstrahlung is considered.[21]

As mentioned before, the energy threshold of the Ge detector will be lowered soon to 1 KeV, embracing the region in which the expected rate is one order of magnitude larger (see Fig. 7). Furthermore the background is projected to be reduced by a factor of 10 or more. This will mean a total improvement of two orders of magnitude, at least, in the bound on the rate which depends on $(g)^4$. Thus the bound $g^{-1} > 1.8{\times}10^7\ GeV$ is expected soon. (For $g^{-1} = 1.8\ 10^7\ GeV$, 23 counts /(kg month) are expected with energy deposition greater than 1 keV). Further improvements are possible in the second generation of ultralow Ge-detectors with a multidetector structure and considerably higher mass expected for 1987-88. One could eventually set limits $g^{-1} > 10^8\ GeV\,(v_T < 0.3\ MeV)$ or, a more exciting possibility, detect solar axions or other light bosons. These bounds are less severe than some more speculative astrophysical bounds which rely, however, on a detailed understanding of the dynamics and evolution of red giants, white dwarfs or other stars very different from the sun. It should be emphasized that the bounds obtained with the Ge detector are laboratory bounds which depend on the knowledge of the best known star, the sun.

Light bosons appear usually only as Nambu-Goldstone bosons (or pseudo N-G bosons). Thus they have pseudoscalar flavor diagonal couplings. Let us mention however for completeness, bounds on light scalars $\tilde{\phi}$ with coupling

$$L = \tilde{g}\ \overline{ee}\,\tilde{\phi}$$

to electrons. The cross-section for "scalarelectric" effect (similar to the previously treated "axiolectric" one) is

$$\sigma_{scalarelectric} \simeq \frac{\tilde{g}^2}{4\pi\alpha_{em}} \sigma_{photoelectric}$$

It does not have the suppression factor $(\omega/2m_e)^2$ thus bounds on $\tilde{g}$ are $\sim 10^3$ times better than bounds on g, for ω around 1 keV. [4]

1. Conclusions

The present low levels of background achieved by the PNL/USC Ge prototype detector represent a new technology which yields interesting bounds on galactic cold dark matter and on light bosons emitted from the sun (besides the purpose for which it was developed i.e., searches for double-β-decay). the main results obtained are presented in Figs. 4 and 7. A reduction of one order of magnitude in the background and the lowering of the energy threshold to 1 keV in initial electron energy are expected soon. Then the rejection or detection of coherently weakly interacting massive particles of mass $m > 8$ GeV as main components of the halo should be possible (unless the R.E.F. becomes very large at some nuclear energy recoil, larger than the new expected threshold). If detection occurs, confirmation will be attempted by looking for an annual modulation. It will be difficult to reduce the energy threshold below 1 keV, thus the detection of particles of lower mass will require other detectors. The Ge-detector is also not sensitive to particles (like the photino) that couple through spin dependent weak interactions.

The two improvements mentioned above should permit reaching the bounds $F/2x_e' > 1.8 \ 10^7$ GeV for axions or $v_T \leq 1.9$ MeV for triplet Majorons (instead of the present $F/2x_e' > 0.5\times10^7$ GeV and $v_T \leq 6.9$ MeV). The present detector is a prototype of 0.710 kg of Ge. Further improvements will be possible with a second generation multicrystal spectrometer (1987-88), in which one ultrapure Ge detector will be surrounded by other Ge detectors acting as an active shield. The PNL/USC groups plan to use 14 detectors of 0.7 kg. With a detection of mass ~ 10 kg a limit of $F/2x_e' \geq 1.0 \ 10^8$ GeV ($v_T \leq 300$ keV) may be achieved, or solar axions or other light bosons could be seen.

Acknowledgements

I thank the participants of the *Quarks and Galaxies* workshop for their questions which lead to the inclusion here of the statement of confidence levels of the bounds on dark matter, and to Figure 7, relative to sneutrinos (which were not present at the time of my talk).

References

1. For a recent review see *Dark Matter in the Universe*, eds. J. Knapp and J. Kormendy (Reidel, 1986), proceedings of the International Astronomical Union Symposium No. 117, Princeton, June 24-28, 1985.

2. For a recent review see P. F. Smith *Possible Experiments for Direct Detection of Particle Candidates for the Galactic Dark Matter*, Rutherford Lab preprint RAL-86-029 to be published in the proceedings of the 2nd ESO/CERN Symposium on *Cosmology, Astronomy and Fundamental Physics*, Garching, March 1986.

3. S. P. Ahlen, F. T. Avignone III, R.L. Brodzinski, A. K. Drukier, G. Gelmini and D. N. Spergel, *Limits on Cold Dark Matter Candidates from the Ultralow germanium Spectrometer*, Harvard Center for Astrophysics Preprint No. 2292, 1986. To appear in Phys. Rev. Lett.

4. F. T. Avignone III, R. L. Brodzinski, S. Dimopoulos, A. K. Drukier, G. Gelmini, B. W. Lynn, D. N. Spergel, G. D. Starkman, *Laboratory Limits on Solar Axions from an Ultralow Background germanium Spectrometer*, SLAC preprint PUB-3872, 1986 to appear in Phys. Rev. D.

5. M. W. Goodman and E. Witten, Phys. Rev. **D31**, 3059 (1985).

6. I. Wasserman, Phys. Rev. **D33**, 2071 (1985).

7. A. K. Drukier, K. Freese, D. N. Spergel, Phys. Rev. D. **33**, 3495 (1986).

8. E. Kolb and M. Turner, Phys. Rev. **D33**, 1202 (1986).

9. J. S. Hagelin, G. L. Kane and S. Raby, Nucl. Phys. **B241**, 638 (1984); L. E. Ibañez Phys. Lett. **137B**, 160 (1984).

10. H. Goldberg, Phys. Rev. Lett. **50**, 1419 (1983); J. Ellis, J. S. Hagelin, D. V. Nanopoulos, K. Olive and M. Srednicki, Nucl. Phys. **B283**, 453 (1984).

11. R. L. Brodzinski, D. P. Brown, J. C. Evans Jr., W. K. Hensley, J. H. Reeves, N. A. Wogman, F. T. Avignone III and H. S. Miley, Nucl. Instr. and Meth. **A239**, 207 (1985).

12. C. Chasman, K. W. Jones and R. A. Ristinen, Phys. Rev. Lett. **15**, 245 (1965).

13. J. Bernstein, L. S. Brown and G. Feinberg, Phys. Rev. **D32**, 3261 (1985).

14. J. Yang, M. S. Turner, G. Steigman, D. N. Schramm and K. A. Olive, Ap. J. **281**, 492 (1984).

15. R. D. Peccei and H. R. Quinn, Phys. Rev. Lett. **3B** (1977) 1440; Phys. Rev. **D16** (1977) 1791. S. Weinberg, Phys. Rev. Lett. **40** (1978) 223. F. Wilczek, Phys. Rev. Lett. **40** (1978) 279. J. E. Kim, Phys. Rev. Lett. **43** (1979) 103. M. A. Shifman, A. I. Vainshtein, and V. I. Zakharov, Nucl. Phys. **B166** (1980) 493. M. Dine, W. Fischler, and M.

Srednicki, Phys. Lett. **104B** (1981) 199. J. E. Moody and F. Wilczek, Phys. Rev. **D30** (1984) 130.

6. D. B. Reiss, Phys. Lett. **115B** (1982) 217; F. Wilczek, Phys. Rev. Lett. **49** (1982) 1549; B. Gelmini, S. Nussinov and T. Yanagida, Nucl. Phys. **B219** (1983) 31.

7. Y. Chicashige, R. N. Mohapatra and R. D. Peccei, Phys. Lett. **98B** (1981) 265, Phys. Rev. Lett. **45** (1980) 1926.

8. G. B. Gelmini and M. Roncadelli, Phys. Lett. **99B** (1981) 411. H. Georgi, S. L. Glashow and S. Nussinov, Nucl. Phys. **B193** (1981) 297.

9. D. S. Dicus, E. W. Kolb, V. L. Teplitz, and R. V. Wagoner, Phys. Rev. **D18** (1978) 1829 and Phys. Rev. **D22** (1980) 829. M. Fukujia, S. Watamura, and M. Yoshimura, Phys. Rev. **48** (1982) 1522.

0. L. M. Krauss, J. E. Moody, and F. Wilczek, Phys. Lett. **B144** (1984) 391.

1. G. G. Raffelt, Phys. Rev. **D33** (1986) 97.

2. S. Dimopoulos, B. W. Lynn and G. D Starkman, Phys. Lett. **168B**, 145 (1986).

LARGE SCALE STREAMING VELOCITIES IN BIASED GALAXY FORMATION

Benjamin Grinstein

California Institute of Technology, Pasadena, CA 91125

ABSTRACT

Even over truly large scales, the velocity of objects whose den—
sity is described by biasing cannot, in general, be accurately calcu—
lated by means of linear perturbation theory. An introductory
account of the methods used to reach this conclusion is given. The
implications are briefly discussed.

1. Introduction and Conclusions

There are two important reasons why the study of the large scale structure
of the universe has received increasing attention by theoretical physicists and
astrophysicists over the last years. First, the body of observational data has
grown and produced interesting information such as the existence of large
voids like the one in Boötes[1] and filamentary—like superstructures such as the
Perseus—Pisces chain[2]. It is probably reasonable to expect to have reliable
measurements of the distribution of galaxies and of clusters of various
richnesses and of their velocities in the near future. Secondly, it has become
apparent that the evolution of structure in the universe is very sensitive to
fundamental physics. As a result, a multitude of speculations from particle
physics has been taken over to cosmology, yielding anything from a catalogue of
candidates for dark matter[3] to loops of string[4] acting as seeds for galaxy for—
mation. Hopefully, understanding of the universe at large will teach us more
about fundamental physics. All of this has been dealt with in other talks at this
workshop, so I will spare the readers from one more lengthy introductory review
and refer them to the rest of the excellent articles in this volume.

There is one other reason why I find the study of large scale structure
appealing. It is reasonable to expect that the effects of gravitational forces on
the evolution of the universe at large are small, and therefore amenable to
analytic treatment through the use of reasonable approximations. The main
subject of this talk is, in fact, an attempt to understand to what extent and
under what conditions one can use linearized perturbation theory[5] in calcu—
lating the effect of gravity on the evolution of the spatial and velocity distribu—
tions of matter; and of the spatial and velocity distributions of regions where
the matter density is unusually high. While our intuition is not likely to fail us
in understanding the gravitational evolution of long wavelength density fluc—

tuations of a fluid on time scales much shorter than it takes for them to collapse, it is not clear that one can easily guess the behavior of statistical distributions of nonlinear functionals of the mass density, (e.g., one which associates a delta function with a local maximum of the filtered density). To address these issues a careful formal treatment is probably required. There is no better way of making this point clear than to state right away the conclusions of this work[6]. Consider* different biasing functionals of the underlying mass density $n_1(\rho(\vec{x}))$, $n_2(\rho(\vec{x}))$, $\cdots$, which are to be interpreted as the number density random field of different types of objects. Assume the velocity of each object is the local velocity of the underlying fluid $\vec{v}(\vec{x})$. Then the velocities $v_i(R)$ of these objects averaged over a fuzzy sphere of radius R become gaussian random variables as $R \to \infty$, so that their distribution is determined by $v_i(R)_{rms}$, but this quantity cannot be reliably computed in the linear approximation. Moreover, if R is large, but not infinite, so that the expectation value of $\vec{v}_i(R)$ does not necessarily vanish, then its expectation value is proportional to that of the underlying velocity field averaged over the same fuzzy ball $\vec{v}(R)$, with the constant of proportionality being precisely $v(R)_{rms}/v_i(R)_{rms}$; the corrections to this statement are order $1/R$. We must make some remarks about this result. First, without more powerful techniques at hand we can presently determine the numbers $v_i(R)_{rms}$ for any specific biasing only through numerical simulations. Second, it is nonetheless encouraging that precise statements can be made from an all orders perturbative analysis. Third, to make this result useful, one must determine how large R has to be before the corrections become small. I will not address this issue here, but the interested reader can find a discussion on the issue in ref. 6.

The linearized theory is a valid approximation at very early times if the initial conditions are only very slightly inhomogeneous. Because fluctuations tend to grow, the approximation becomes poor at later times. A "time expansion" requires that we keep more terms if we expect it to remain valid at later times. Similarly, if we average over initial conditions (i.e., if we take the initial perturbations to be a gaussian random field), the evolution of the mean of any quantity will be poorly described by linear perturbation theory at late times. Assume, though, that the quantity of interest is a power of the filtered density, where the filtering scale R is large. Then there is a systematic expansion in inverse powers of R. This expansion is valid when the linear approximation (retaining only leading powers of time) has broken down. Each order in $1/R$ may contain contributions from all orders in the "time expansion." Granted, this makes it hard to compute anything, but that we can still learn a great deal is a welcome surprise. This is discussed in section 2. A similar conclusion can be drawn for correlations of fluctuations at points separated by a large distance R. Section 3 describes how to derive the conclusion described in the preceding paragraphs, while section 4 contains a brief discussion of the results.

2. Method and Preliminary Results

We will work in the context of an $\Omega = 1$ universe with cold dark matter, and assume that the mass can be described as a newtonian pressureless fluid. The

---

* We work in an $\Omega = 1$ universe, with adiabatic cold dark matter and a Zeldovich spectrum of perturbations.

Robertson–Walker factor is* $R \sim t^{2/3}$ and the background density $\overline{\rho} \sim R^{-3} \sim t^{-2}$. The fluid is completely described by specifying its density contrast with respect to the background $\delta(\vec{x}, t) = \dfrac{\rho(\vec{x}, t)}{\overline{\rho}} - 1$ and its peculiar velocity $\vec{v}(\vec{x}, t)$, both as functions of the comoving spatial coordinate $\vec{x}$ and the time t. The Navier–Stokes and the continuity equations give the evolution of these fields:

$$\frac{\partial \delta}{\partial t} + \frac{1}{R} \vec{\nabla} \cdot (1 + \delta)\vec{v} = 0 \tag{1a}$$

$$\frac{\partial \vec{v}}{\partial t} + \frac{\dot{R}}{R} \vec{v} + \frac{1}{R}(\vec{v} \cdot \vec{\nabla})\vec{v} = \vec{g} \tag{1b}$$

$$\vec{\nabla} \cdot \vec{g} = -4\pi G\, \overline{\rho} R\, \delta, \quad \vec{\nabla} \times \vec{g} = 0. \tag{1c}$$

Because they are nonlinear, these equations are difficult to solve for arbitrary initial conditions. If the initial condition has both δ and $\vec{v}$ small, say order ε, one may try to find an approximate solution by expanding in powers of ε. The leading approximation $\delta_1(\vec{x}, t) \sim O(\varepsilon)$ is obtained by solving the eqs. (1) when all nonlinear terms are dropped. Of the two independent solutions only one grows with time $(\sim t^{2/3})$ so we discard the dying one $(\sim t^{-1/3})$. The leading solution for $\vec{v}_1$ is totally determined in terms of δ_1. The next order gives a correction $\delta_2(\vec{x}, t) \sim O(\varepsilon^2)$ which satisfies an equation obtained from (1) by substituting $\delta_1 + \delta_2$ for δ, $\vec{v}_1 + \vec{v}_2$ for $\vec{v}$ and dropping terms of $O(\varepsilon^3)$. This determines $\delta_2(\vec{x}, t)$ and $\vec{v}_2(\vec{x}, t)$ in terms of $\delta_1(\vec{x}, t)$. Both are quadratic in δ_1, and δ_2 is $\sim (\varepsilon\, t^{2/3})^2$. Clearly we could iterate this procedure and obtain higher order corrections** but in order to illustrate our methods it will suffice to include only this first nonlinear term. The thing to note is that there is some computable function $P_2(\vec{y}, \vec{z})$ such that, if $\delta_1(\vec{x}, t) = \varepsilon_1(\vec{x})t^{2/3}$, then $\delta_2(\vec{x}, t) = \varepsilon_2(\vec{x})t^{4/3}$ where

$$\varepsilon_2(\vec{x}) = \int d^3y\; d^3z\, P_2(\vec{x}-\vec{y}, \vec{x}-\vec{z})\, \varepsilon_1(\vec{y})\, \varepsilon_1(\vec{z}). \tag{2}$$

We can make use of this result immediately. Assume, as usual, that $\varepsilon_1(\vec{x})$ is a gaussian random field, so that it has no connected (reduced) correlations other than the "two-point" function $\langle \varepsilon_1(\vec{x})\varepsilon_1(\vec{y}) \rangle$. Then at very early times $\delta(\vec{x}, t)$ is also a gaussian random field, but later the $\delta_2(\vec{x}, t)$ piece will induce higher connected correlations. To the order we are working we have, for example

$$\langle \delta(\vec{x}, t)\delta(\vec{x}_2, t)\delta(\vec{x}_3, t) \rangle = t^{8/3}(\langle \varepsilon_2(\vec{x}_1)\varepsilon_1(\vec{x}_2)\varepsilon_1(\vec{x}_3) \rangle$$
$$+ \text{permutations}) + O(t^{10/3}) \tag{3}$$

and using (2) we can evaluate

$$\langle \varepsilon_2(\vec{x}_1)\varepsilon_1(\vec{x}_2)\varepsilon_1(\vec{x}_3) \rangle = \int d^3y\; d^3z\, P_2(\vec{x}_1-\vec{y}, \vec{x}_1-\vec{z})$$

$$\langle \varepsilon_1(\vec{y})\varepsilon_1(\vec{z})\varepsilon_1(\vec{x}_2)\varepsilon_1(\vec{x}_3) \rangle$$

$$= \int d^3y\; d^3z\, P_2(\vec{x}_1-\vec{y}, \vec{x}_1-\vec{z})$$

$$[\langle \varepsilon_1(\vec{y})\varepsilon_1(\vec{z}) \rangle\langle \varepsilon_1(\vec{x}_2)\varepsilon_1(\vec{x}_3) \rangle$$

$$+ 2\langle \varepsilon_1(\vec{y})\varepsilon_1(\vec{x}_2) \rangle\langle \varepsilon_1(\vec{z})\varepsilon_1(\vec{x}_3) \rangle]. \tag{4}$$

---

* The reader will excuse me for using the same symbol for both the radius of the sphere and the Robertson–Walker scale factor. Which is being referred to should be clear from the context.
** And this is done in ref. 7.

It is clear that:

- In order to understand the behavior of the three point function that we have just computed we need to understand the function $P_2(\vec{y}, \vec{z})$.

- The calculation of $\left\langle \delta(\vec{x}_1, t)\delta(\vec{x}_2, t) \cdots \delta(\vec{x}_{17}, t)\right\rangle$ is straightforward but very lengthy and tedious: one is bound to make mistakes unless there is a simple way to systematically carry it out.

- We could have kept higher orders in our expansion of $\delta(\vec{x}, t)$, but the calculation of connected correlations would have been unaltered in essence. Functions $P_3(\vec{x}_1, \vec{x}_2, \vec{x}_3)$, $P_4(\vec{x}_1, \cdots, \vec{x}_4)$, etc., would have been introduced and more terms would have had to be considered.

We will address the first issue later. For now we concentrate on the last two. Clearly what we need is a bookkeeping procedure that will make computations tractable*.

A simple way of accomplishing this is by introducing a diagrammatic notation, as follows:

Diagrams with their corresponding equations and expressions:

$$\varepsilon_2(\vec{x}) = \int d^3y \, d^3z \, P_2(\vec{x}-\vec{y}, \vec{x}-\vec{z})\varepsilon_1(\vec{y})\,\varepsilon_1(\vec{z})$$

$$\varepsilon_1(\vec{x}) = \int d^3y \, P_1(\vec{x}-\vec{y})\,\varepsilon_1(\vec{y})$$

$$\left\langle \varepsilon_1(\vec{x})\,\varepsilon_1(\vec{y})\right\rangle \tag{5}$$

We have introduced $P_1(\vec{x}-\vec{y}) = \delta^{(3)}(\vec{x}-\vec{y})$ since it is convenient to have external points represented only by bold lines. For example the computation in (4) is represented by

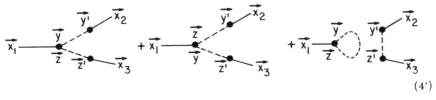

$$\tag{4'}$$

---

* We will settle for one that will make the analysis of the behavior of correlation functions feasible.

The expressions corresponding to these diagrams are easily reconstructed using the above rules, so that one may classify all diagrams before actually performing a calculation. The generalization to diagrams involving ε_3, ε_4, etc. is straightforward. Notice that it is trivial to see which diagrams contribute to connected correlations. This is only a partial solution to the two issues raised above since there is still some bookkeeping left to be done associated with the classification of diagrams that contribute to a given correlation to any given order in $\varepsilon\, t^{2/3}$, e.g., with going from eq. (3) to eq. (4) in our example.

Another useful exercise is to consider the correlations of the density contrast averaged over a ball of radius R:

$$\delta(R,t) \equiv \frac{1}{R^3} \int d^3x\, W_R(x)\, \delta(\vec{x}, t). \tag{6}$$

Here $W_R(x)$ is a window function of characteristic scale R. We may take $W_R(x) = e^{-x^2/R^2}$ as a concrete example to keep in mind. From our discussion above, there is a contribution to $\langle \delta(R,t)^3 \rangle_{\text{connected}}$ that goes like (up to a counting factor)

$$t^{8/3}\frac{1}{R^9} \int d^3x_1\, d^3x_2\, d^3x_3\, W_R(x_1) W_R(x_2) W_R(x_3)$$

$$\times \int d^3y\, d^3z\, P_2(\vec{x}_1 - \vec{y}, \vec{x}_1 - \vec{z}) \langle \varepsilon_1(\vec{y})\, \varepsilon_1(\vec{x}_2) \rangle \langle (\varepsilon_1(\vec{z})\, \varepsilon_1(\vec{x}_3) \rangle. \tag{7}$$

If t is not small, we may want to compare this to contributions of order $t^{12/3}$, such as the one from the diagram

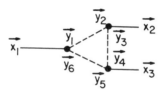

that is

$$t^{12/3}\frac{1}{R^9} \int d^3x_1\, d^3x_2\, d^3x_3 W_R(x_1) W_R(x_2) W_R(x_3)$$

$$\times \int (\prod_{i=1}^{6}) P_2(\vec{x}_1 - \vec{y}_1, \vec{x}_2 - \vec{y}_6) P_2(\vec{x}_2 - \vec{y}_2, \vec{x}_2 - \vec{y}_3) P_2(\vec{x}_3 - \vec{y}_4, \vec{x}_3 - \vec{y}_5)$$

$$\times \langle \varepsilon_1(\vec{y}_1)\, \varepsilon_1(\vec{y}_2) \rangle \langle \varepsilon_1(\vec{y}_3)\, \varepsilon_1(\vec{y}_4) \rangle \langle \varepsilon_1(\vec{y}_5)\, \varepsilon_1(\vec{y}_6) \rangle. \tag{8}$$

We have not included combinatorial factors in (7) and (8) since we want to compare these expressions <u>as functions of R</u>, at least for large R. Clearly we need at this point some more information about the two-point function and the function P_2. For this purpose it is convenient to introduce their Fourier transforms

$$\varepsilon_n(\vec{k}) = \int d^3x\, e^{i\vec{k}\cdot\vec{x}} \varepsilon_n(\vec{x})$$

$$W_R(k) = \int d^3x\, e^{i\vec{k}\cdot\vec{x}} W_R(x)$$

$$\langle \varepsilon_1(\vec{x})\, \varepsilon_1(\vec{y}) \rangle = \int \frac{d^3k}{(2\pi)^3}\, e^{-i\vec{k}\cdot(\vec{x}-\vec{y})} P(k)$$

$$P_2(\vec{k}_1, \vec{k}_2) = \int d^3x_1\, d^3x_2\, e^{i\vec{k}_1\cdot\vec{x}_1 + i\vec{k}_2\cdot\vec{x}_2} P_2(\vec{x}_1, \vec{x}_2).$$

In terms of these we have, for example

$$\varepsilon_2(\vec{k}) = \int \frac{d^3q_1}{(2\pi)^3} \frac{d^3q_2}{(2\pi)^3} (2\pi^3)\delta^{(3)}(\vec{k}-\vec{q}_1-\vec{q}_2)P_2(\vec{q}_1,\vec{q}_2)\varepsilon_1(\vec{q}_1)\varepsilon_1(\vec{q}_2)$$

and

$$\left\langle \varepsilon_1(\vec{k})\varepsilon_1(\vec{k}')\right\rangle = (2\pi)^3\delta^{(3)}(\vec{k}+\vec{k}')P(k).$$

In particular expressions (7) and (8) become, respectively,

$$t^{8/3}\frac{1}{R^9}\int d^3k_1\,d^3k_2\,W_R(\,|\vec{k}_1+\vec{k}_2|\,)W_R(k_1)W_R(k_2)P(k_1)P(k_2)P_2(\vec{k}_1,\vec{k}_2)$$

$$(7')$$

and

$$t^{12/3}\frac{1}{R^9}\int d^3k_1\,d^3k_2\,d^3q\,W_R(\,|\vec{k}_1+\vec{k}_2|\,)W_R(k_1)W_R(k_2)$$

$$P(q)P(\,|\vec{q}-\vec{k}_1|\,)P(\,|\vec{q}+\vec{k}_2|\,)$$

$$P_2(\vec{q}+\vec{k}_2,-\vec{q}+\vec{k}_1)P_2(\vec{q}-\vec{k}_1,-\vec{q})P_2(\vec{q},-\vec{q}-\vec{k}_2).\qquad(8')$$

The interested reader will find it useful to rewrite the diagrammatic rules directly in terms of the Fourier transforms.

We are ready to discuss the properties of P_2 that are needed to compare (7) and (8). We could at this point write the explicit expression for P_2, but this would not be very useful when we attempt to understand P_3, P_4, etc. It is therefore best to list the useful general properties of all the P_n:

1. $P_n(\vec{k}_1, \cdots, \vec{k}_n)$ is rotationally invariant and symmetric under permutation of its arguments.

2. $P_n(\vec{k}_1, \cdots, \vec{k}_n)$ is a homogeneous function of degree zero: for any λ, $P_n(\lambda\vec{k}_1, \cdots, \lambda\vec{k}_n) = P_n(\vec{k}_1, \cdots, \vec{k}_n)$.

3. For nonvanishing $\vec{k}_i$, $P_n(\vec{k}_1, \cdots, \vec{k}_n)$ vanishes quadratically as $\vec{k}_1 + \cdots + \vec{k}_n \to 0$.

4. As $\vec{k}_1 \to 0$, with $\vec{k}_2, \cdots, \vec{k}_n$ fixed, $P_n(\vec{k}_1, \cdots, \vec{k}_n)$ diverges at most linearly, that is $P_n \sim 1/k_1$.

5. Setting $\vec{k}_1 = -\vec{k}_2 = \vec{q}$, letting $q \to \infty$ while $\vec{k}_3, \cdots, \vec{k}_n$ are held fixed, $P_n(\vec{q}, -\vec{q}, \vec{k}_3, \cdots, \vec{k}_n)$ vanishes quadratically with q $(P_n \sim 1/q^2)$. This can be generalized to the case when $\sum_{i=1}^{L<n}\vec{k}_i = 0$, with $k_i \to \infty$ for $i = 1, \cdots, L$ while $\vec{k}_{L+1}, \cdots, \vec{k}_n$ are held fixed, in which case P_n vanishes at least linearly. This is trivial for P_2 since $P_2(\vec{q}, -\vec{q}) = 0$ by property 3, but it is useful in analyzing P_n for $n > 2$.

Property 3 follows from momentum conservation, while property 4 is a kinematic effect and is true even if gravity is switched off (the continuity equation (1a) gives an expression for $\vec{\nabla} \cdot \vec{v}$). The 5th property states that the effect of a very short wavelength fluctuation $A\cos(\vec{q}\cdot\vec{x})$ of fixed amplitude A superimposed on $\varepsilon_1(\vec{x})$ on higher modes filtered over scales larger than $1/q$ is negligible.

The Zeldovich power spectrum for adiabatic cold dark matter has the asymptotic behavior[8]

$$P(k) \sim \begin{cases} k & \text{as } k \to 0 \\ k^{-3}(lnk)^2 \text{ as } k \to \infty. \end{cases} \tag{9}$$

We can finally analyze the R behavior of (7') and (8'). As R becomes large the window functions W_R restrict the region of integration to $k_i < 1/R$. In connection with (7') we are led to investigate the behavior of $P(k_1)P(k_2)P_2(\vec{k}_1, \vec{k}_2)$ when k_1 and k_2 are small. By property 2, when $k_i \to 0$, $P_2(k_1, k_2)$ is of order 1 and $P(k_1)P(k_2)P_2(\vec{k}_1, \vec{k}_2) \sim k_1 k_2 \sim 1/R^2$. The situation is quite different for (8'). We need to analyze

$$\int d^3q \, P(q)P(|\vec{q} - \vec{k}_1|)P(|\vec{q} + \vec{k}_2|)P_2(\vec{q} + \vec{k}_2, -\vec{q} + \vec{k}_1)$$
$$P_2(\vec{q} - \vec{k}_1, -\vec{q})P_2(\vec{q}, -\vec{q} - \vec{k}_2).$$

By property 3 if $k_i \to 0$ holding q fixed, the term $P_2(\vec{q} - \vec{k}_1, -\vec{q})$ gives a factor of order k_1^2, $P_2(\vec{q}, -\vec{q} - \vec{k}_2)$ gives a factor of k_2^2 and $P_2(\vec{q} + \vec{k}_2, -\vec{q} + \vec{k}_1) \sim (\vec{k}_1 + \vec{k}_2)^2$. Therefore the expression (8') picks up a factor of $1/R^6$. Thus expression (8) vanishes more rapidly than (7) by a factor of $1/R^4$ as R becomes large. Please note that the asymptotic form of the power spectrum (9) ensures convergence of the integration over $\vec{q}$.

No use was made of properties 4 and 5 in reaching the above conclusion. This is merely a reflection of the fact that the case we considered is simpler than the general one. Still, one general feature emerges in our examples: the "tree" graph dominates over "loop" graph as R gets large. In fact this situation persists to all orders. Using properties 1–5 it has been proved that under our assumptions $\langle \delta(R, t)^n \rangle$ is dominated for large R by the contributions arising from tree diagrams. An analogous conclusion holds when considering $\langle v(R, t)^m \delta(R, t)^n \rangle$, where $\vec{v}(R, t)$ is the velocity field averaged over a ball, defined as in (6).

Before leaving this section we give one more simple example. Consider the effects of nonlinear evolution on $\langle \delta(\vec{x}, t)\delta(\vec{y}, t) \rangle$. Typical diagrams are

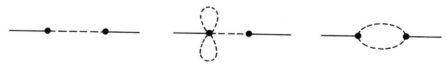

The reader should find no trouble in checking that the Fourier transforms of these behave as k, k^3 and k^4 respectively, as $k \to 0$. For a general power spectrum the results would be $P(k)$, $k^2 P(k)$ and k^4, respectively. This celebrated result[5] states that even if the power spectrum initially has no power at long distances (small k), it develops a k^4 tail through non–linear mode–coupling effects. This is but one example of a situation where short–distance fluctuations can give rise to long distance correlations because of gravitational interactions. Notice that the magnitude of such an effect is noncomputable (in "time" perturbation theory) in the sense that diagrams such as

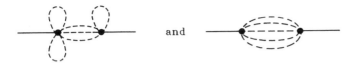

and

are also order k^4 as $k \to 0$.

3. Velocities

Our methods can be applied readily to the study of the distribution of velocities. In particular, we have already seen that $\langle v(R,t)^2 \rangle$ is dominated by the "tree" diagram contribution, which for a two-point function is just the statement that $\langle v(R,t)^2 \rangle$ is well approximated by the linear theory. Therefore knowledge of $\langle v(R,t)^2 \rangle$ yields direct information about the primordial power spectrum $P(k)$.

Unfortunately there is no way of measuring $\langle v(R,t)^2 \rangle$ directly, at least if indeed $\Omega = 1$ so that most matter is dark. For then our measurement of velocities is biased by our ability to perform it only at very special points, namely those where luminous objects have formed. There is no reason a priori to expect that the velocities of galaxies constitute a fair sample for the velocity of the underlying fluid $\vec{v}(\vec{x}, t)$, even if the galaxies are moving with the fluid's local velocity. A quantity more easily related to observation is the autocorrelation of the product $n_i(\vec{x}, t)\,\vec{v}(\vec{x}, t)$ where $n_i(\vec{x}, t)$ is the spatial density of objects of type i. Thus we are led to define[*]

$$\vec{u}_i(R,t) \equiv \frac{1}{\bar{n}_i} \int d^3x \, n_i(\vec{x}, t)\, \vec{v}(\vec{x}, t)\, W_R(x),$$

and to consider $\langle \, | \vec{u}_i(R,t) |^2 \rangle$ or more generally $\langle \vec{u}_i(R,t) \cdot \vec{u}_j(R,t) \rangle$. Expanding we have

$$\langle \vec{u}_i(R,t) \cdot \vec{u}_j(R,t) \rangle = \frac{1}{\bar{n}_i \bar{n}_j} \int d^3x \, d^3x' \, W_R(x)\, W_R(x')$$

$$\langle n_i(\vec{x}, t)\, \vec{v}(\vec{x}, t) \cdot n_j(\vec{x}', t)\, \vec{v}(\vec{x}', t) \rangle. \tag{10}$$

In studying the autocorrelation of $n_i(\vec{x}, t)\, \vec{v}(\vec{x}, t)$ it will be convenient to write $\Delta_i = \dfrac{n_i}{\bar{n}_i} - 1$ so that one can see more readily whether the expectation value of $u(R)^2$ differs significantly from that of $v(R)^2$:

$$\langle u_i(R,t)^2 \rangle - \langle v(R,t)^2 \rangle = \int d^3x \, d^3x' \, W_R(x)\, W_R(x')$$

$$\langle (2 + \Delta(\vec{x}, t))\vec{v}(\vec{x}, t) \cdot \Delta(\vec{x}', t)\, \vec{v}(\vec{x}', t) \rangle.$$

To go any further we need information about the distribution of the density $n_i(\vec{x}, t)$ and its correlation to $\vec{v}(\vec{x}, t)$. While we do not understand how objects form, it is sensible to imagine that their density can be given in principle as some functional of the density contrast $n_i(\vec{x}) = n_i(\delta(\vec{x}))$, at least in the sense that correlations of $n_i(\vec{x})$ can be calculated in terms of those of $\delta(\vec{x})$ through

---

[*] A somewhat better definition is given in ref. [6].

the functional $n_i(\delta(\vec{x}))$. This is commonly referred to as "biasing," and we adopt this procedure in what follows. The reader may want to keep some simple examples in mind. Kaiser[9] introduced an exponential biasing

$$n(\vec{x}, t) \sim e^{T\int d^3y W_{R'}(\vec{x}-\vec{y})\delta(\vec{y},t)}.$$ (11)

There is also points above a threshold[10),11)

$$n(\vec{x}, t) \sim \theta(\int d^3y \, W_R(\vec{x} - \vec{y})\delta(\vec{y}, t) - T),$$ (12)

where θ is a step-function. For illustration it will be convenient to have in mind the simplest biasing ("objects tracing the mass")

$$n(\vec{x}, t) = \overline{n}(1 + b\delta(\vec{x}, t)),$$ (13)

and the simplest nonlinear biasing

$$n(\vec{x}, t) = \overline{n}[1 + b\delta(\vec{x}, t) + a(\delta(\vec{x}, t)^2 - \langle \delta(\vec{x}, t)^2 \rangle)].$$ (14)

Here a and b are small arbitrary constants.

We are now ready to compare the expectation values of $u(R)^2$ and of $v(R)^2$. Let us take the simplest example, that of objects tracing the mass (13). Let us compare the behavior of $\langle \vec{v}(\vec{x}, t) \cdot \vec{v}(\vec{x}', t) \rangle$ and $\langle \vec{v}(\vec{x}, t) \cdot \delta(\vec{x}', t)\vec{v}(\vec{x}', t) \rangle$ as $|\vec{x} - \vec{x}'|$ gets large. Some diagrams contributing to the former are

Here, the wavy line stands for the velocity field. As we have seen, for a Zeldovich spectrum the first diagram dominates at long distances, and this persists to all orders (loop diagrams are subdominant). For the second correlation, the diagrams that contribute to first order in the perturbative expansion (i.e., those which dominate at early times) are

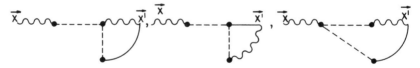

The last diagram is subdominant for large $|\vec{x} - \vec{x}'|$, as compared with the first two. This is most easily seen in Fourier space by counting powers of k. There is only one variable k since by translational invariance the result can be written as $\int d^3k \, e^{i\vec{k}\cdot(\vec{x}-\vec{x}')}f(k)$. The feature that makes the first two diagrams dominant is not the absence of loops but that there is a line joining both sides which carries the "external" wavenumber $\vec{k}$. Counting powers of k we see that each of these diagrams scales at long distances just as the leading contribution to $\langle \vec{v}(\vec{x}, t) \cdot \vec{v}(\vec{x}', t) \rangle$. This feature will persist for other biasing functionals. Still, something remarkable happens when we calculate the diagrams explicitly: in their sum the leading term as $k \to 0$ cancels! This persists to all orders and occurs also in the calculation of $\langle \delta(\vec{x}, t)\vec{v}(\vec{x}, t) \cdot \delta(\vec{x}', t)\vec{v}(\vec{x}', t) \rangle$. These state-ments are proved in ref. [6] by noting that the correlations of $\delta(\vec{x}, t)\vec{v}(\vec{x}, t)$ are related to those of $\delta(\vec{x}, t)$ alone by the equation of continuity (1a). Therefore, for objects that trace the mass $u(R, t)_{rms} = v(R, t)_{rms}(1 + O(1/R))$.

The above exercise teaches us how to characterize the diagrams which give corrections to $u(R,t)_{rms}$ which scale at large R just like $v(R,t)_{rms}$ does. It also shows that all such corrections may just cancel, so that care must be used in reaching any definite conclusions. Nevertheless, since these cancellations could be traced back to the equation of continuity, which could be used only because of the special form of that particular biasing, we do not expect a priori that they will operate for other more complicated functionals, such as (11) or (12). In fact it is very easy to check that the leading contributions to $\langle \vec{v}(\vec{x},t) \cdot \Delta(\vec{x}',t)\vec{v}(\vec{x}',t) \rangle$ for the quadratic biasing (14) arise from the graphs

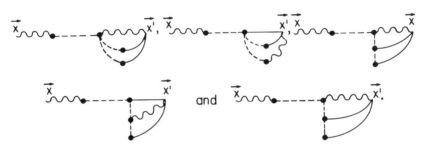

Moreover, they each scale with $|\vec{x} - \vec{x}'|$ just as $\langle \vec{v}(\vec{x},t) \cdot \vec{v}(\vec{x}',t) \rangle$ and do not cancel in the sum. The leading contributions to $\langle \Delta(\vec{x},t)\,\vec{v}(\vec{x},t) \cdot \Delta(\vec{x}',t)\,\vec{v}(\vec{x}',t) \rangle$ in this case correspond to graphs obtained by joining in all possible ways the right halves of the figures above. For example

These again scale at large distances like the ones above. Therefore $u(R,t)_{rms}$ does not generally equal $v(R,t)_{rms}$. The precise relation requires a computation of $\vec{v} \cdot \Delta'\vec{v}'$ and $\Delta\vec{v} \cdot \Delta'\vec{v}'$ to all orders in our expansion and therefore cannot be obtained. Still, the dominant diagrams have a common feature: they are of the form

they can be divided into two pieces by cutting a single internal line. Even if we cannot compute the relation between $u(R,t)_{rms}$ and $v(R,t)_{rms}$, we can still extract some information. The leading k dependence as $k \to 0$ in (15) is given by property 4 in the previous section: $\sim \frac{1}{k} P(k) \frac{1}{k}$. Thus, as function of R, the velocities are proportional

$$u_i(R,t)_{rms} = c_i(t)v(R,t)_{rms}(1 + O(1/R)).$$

This is our main result. We have included here a reference to the type of object, since the argument clearly applies to any type of biasing.

4. Final Comments

A bit more work shows that the above holds at the level of vectors[6]

$$\vec{u}_i(R,t) = c_i(t)\,\vec{v}(R,t)(1 + O(1/R)).$$

The burning issue is to find out what the constants c_i are for particular biasing functionals. As mentioned before, they are not analytically calculable with present day techniques. One must resort to numerical simulations. Special care must be used in performing numerical studies aimed at understanding velocities. Since the effect described here arises strictly from the coupling of short and long wavelength modes, the simulations need to represent the physics at short as well as long distances fairly. Unfortunately present studies often introduce a softening procedure[12] which smooths out short distances and so the physics at the smaller scales may not be well represented.

In the limit of large threshold (the parameter T in (11) or (12)), keeping the normalization of $\langle n_i(\vec{x},t)n_i(\vec{x}',t)\rangle$ fixed, it is not difficult to see that c_i-1 is of order $1/T$. If for reasonable (?) choices of the biasing functional the constants c_i are always found to be close to or smaller than unity, then one may argue that the model ($\Omega = 1$ adiabatic cold dark matter with gaussian Zeldovich perturbations) confronts observations unfavorably.

REFERENCES

1. Kirschner, R.P., Oemler, A. Jr., Schecter, P.L., and Schectman, S.A., Ap. J. Letters, 248 (1981) L57.

2. Gregory, S.A., Thompson, L.A., and Tifft, W.G., Ap. J., 243 (1981) 411.

3. See for example Olive, K., "Dark Matter: Cosmological limits and applications" FERMILAB−CONF−85/123−A(1985).

4. Albrecht, A., and Turok, N., Phys. Rev. Lett. 54 (1985) 1868. For a recent review see Vilenkin, A., Phys. Rep. 121 (1985) 263.

5. A good introduction to the subject can be found in Peebles, P.J.E., "The large scale structure of the universe," Princeton University Press, Princeton, NJ, 1980.

6. Grinstein, B., Politzer, H.D., Rey, S.−J., and Wise, M.B., "Interpretation of large scale deviations from the Hubble flow," CALT−68−1358.

7. Goroff, M.H., Grinstein, B., Rey, S.−J., and Wise, M.B., "Coupling of modes of cosmological mass density fluctuations," CALT−68−1342.

8. Bardeen, J.M., Bond, J.R., Kaiser, N., and Szalay, A.S., Ap. J., 304 (1986) 15.

9. Kaiser, N., Ap. J. Letters 284 (1984) L9.

10. Kaiser, N., in Proc. Inner Space/Outer Space Conference, ed., Kolb, E. W., Turner, M. S., Olive, K., Seckel, D., and Lindley, D., University of Chicago Press, Chicago 1985.

11. Politzer, H.D., and Wise, M.B., Ap. J. Lett., 285 (1984) L1.

12. White, S.D.M., Frenk, C.S., Davis, M., and Efstathiou, G., Steward Observatory preprint No. 666.

DOES THE PHOTINO DECAY?

Lawrence J. Hall

Lawrence Berkeley Laboratory
and
Department of Physics
University of California
Berkeley, California 94720

Abstract

The stability of the lightest superpartner is a crucial aspect of many experimental searches for supersymmetry and of supersymmetric dark matter candidates. It is shown that R parity may occur in operators of dimension four or less as an accidental consequence of an exact Z_N symmetry. In this case the lightest superpartner can decay via higher dimension operators. The lifetime depends on the scale of the new physics responsible for the non-renormalizable operators; it could be anywhere in the region 10^{-20} seconds to 10^{+20} seconds. Explicit examples are given.

Particle physicists believe that the standard model must be incorporated in a more unified framework because there is so much which the standard model cannot explain. This is an oft-quoted motivation for studying extensions of the standard model, such as supersymmetric theories. However, in the standard model there are some things that we really do understand: the proton is stable and the neutrino is massless because there are no gauge invariant, renormalizable operators which violate baryon (B) and lepton (L) number. The large number of searches for proton decay and for neutrino masses are due to the widespread belief that the standard model cannot be the whole story. The negative results to date are an indication that the standard model, at least in some areas, is a fantastically good approximation to the ultimate theory.

It is astonishing that supersymmetric theories are so popular, given that in the supersymmetric version of the standard model it is no longer possible to understand the conservation of B and L as a result of gauge invariance and renormalizability. These symmetries could be broken by

$$(\ell h_1, \ell\ell e^c, q\ell d^c, u^c d^c d^c]_F, \qquad (1)$$

where q, ℓ are SU(2) doublet quark and lepton superfields, u^c, d^c, e^c are singlet antiquark and antilepton superfields, and h_1, h_2 are Higgs doublet superfields.

Why should the coefficients of the operators be zero, or at least small? One possibility is for reasons of chiral symmetry: they are small, perhaps $\sim 10^{-5}$, for the same reason that the electron Yukawa coupling is small. This is fine for the lepton number violating operators, but not for the operators $u^c d^c d^c$.

Another possibility is that the theory possesses a Z_2 symmetry under which all quark and lepton superfields change sign. This matter parity, $\dot{M}_2$, forbids the operators (1). In fact, we can define matter and Higgs fields by whether they are $-$ or $+$ under M_2. Even if we introduce exotic fields with $M_2 +$ or $-$, providing M_2 is conserved, then baryon number is necessarily conserved by renormalizable operators. Indeed M_2 is just

$$M_2 = (-)^{Q+L}, \qquad (2)$$

where Q is quark number and L is lepton number.

Higher dimension operators can violate B and L, even if they conserve M_2, for example

$$[qqq\ell, u^c u^c d^c e^c]_F. \qquad (3)$$

As long as M_2 is conserved, so is R-parity

$$R_2 = M_2(-1)^F, \qquad (4)$$

where F is fermion number. R_2 divides all particles into two sets: familiar particles with R_2 $(+)$ and superpartners with R_2 $(-)$. A theory with M_2 exact is therefore seen to necessarily have a new stable particle: the lightest superpartner (LSP). This is often taken to be the photino. It is worth stressing that if the higher dimension operators preserve M_2, then the LSP is absolutely stable.[1]

The above argument, starting from the observed conservation of B and L and ending with exact stability of the LSP, has become so familiar that it is now generally believed. The LSP is considered to be a likely candidate for the dark matter of the universe. In particle physics experiments, the crucial signature for supersymmetry is considered to be missing energy or missing transverse momenta, which is carried off by the stable neutral LSP.

In this talk I will reexamine the question "Does the photino decay?" Both cosmologists and particle physicists can agree that this is about the most important question of low energy supersymmetry. I will argue that the conventional argument for the stability of the photino is on shaky ground and certainly incomplete. I find it quite plausible that the LSP will decay in cosmological times, so it can be charged; perhaps it is a squark or slepton. It seems probably that some sign of its stability will be apparent in high energy physics collisions, although this is not necessarily true.

The idea that R_2 may be broken is not new: both explicit[2-5] and spontaneous breakings[6,7] have been discussed. With explicit breaking, it has been assumed that some of the operators (1) may be present but that others are absent. This can be due to a combination of global symmetries and accidents. Although this R_2 breaking is viable, the unified schemes which lead to it are ugly.

R_2 may be spontaneously broken via a sneutrino vev. If lepton number L were a good global symmetry, the resulting goldstone boson would cool stars too rapidly unless $\langle \tilde{\nu} \rangle \lesssim 10$ keV. It would therefore seem best to have L broken explicitly, e.g. by neutrino masses, without explicitly breaking R_2. In this case, it has been claimed that large sneutrinos vevs are possible: $\langle \tilde{\nu}_\tau \rangle \sim 10$ GeV. However, in this case the two vacua $\langle \tilde{\nu}_\tau \rangle = \pm$ v are separated by a potential barrier which gives rise to a domain wall.[8] To reduce the energy per unit area of the domain wall to acceptable levels requires: $m^2 v^2 < (10^{-2} \text{ GeV})^4$, where m is the mass of the scalar which would be a goldstone boson in the limit of L conservation. For $m \gtrsim$ MeV, $v \lesssim 100$ MeV. It therefore seems that a new scale of symmetry breaking must be understood in schemes with R_2 broken spontaneously.

In this talk I will consider explicit rather than spontaneous breaking of R_2. However, the explicit violations will be only via operators of dimension five or more. I will show how to justify this new way of breaking R_2 in terms of exact discrete symmetries of the theory.

What is the origin of M_2? It could be that M_2 is an automatic consequence of the structure of a more unified gauge model. For example, in SO(10) models it could be because all interactions have an even number of spinor fields. In E_6

models it could even be a discrete subgroup of the gauge group.[9] It is not even necessary to go to such highly unified schemes: M_2 arises naturally in most left-right symmetric models, where it can be identified as $(-1)^T$ where T is the total number of SU(2) indices. Quark and leptons all have a single SU(2) index, either left or right, and are M_2 $(-)$, while Higgs have both a left and right index and are M_2 $(+)$. M_2 was also automatic in once-popular models with an additional U(1) gauge symmetry. Such examples have given great credibility to M_2, and enhanced the expectation that the LSP is stable.

It is worth looking at the left-right example more closely. It is clear that it is SU(2)$_R$ which forbids operators (1), they all have an odd number of SU(2)$_R$ doublets. Thus the basic origin of B and L conservation is the gauge symmetry. In fact, M_2 is a discrete Z_2 subgroup of the gauge group

$$M_2 = (-1)^{2(T_{3L}+T_{3R})}. \tag{5}$$

Next consider a generalization of matter parity. Suppose the operators (1) are forbidden by a discrete symmetry X which does not contain M_2. Although X and M_2 have the same effect on the theory up to terms of dimension 4, they may differ in the terms which they allow at higher dimension. Thus, while M_2 forbids LSP decay to all orders, X may allow LSP decay. In this talk I will take X to be Z_N, $N > 2$. In superstring inspired models, Z_N symmetries frequently occur from the six dimensional compact manifold. Explicit examples include $Z_5 \times Z_5$[10] and $Z_3 \times Z_3 \times Z_2$.[11]

For simplicity we start our systematic analysis by choosing the low energy gauge group to be SU(3) $\times$ SU(2) $\times$ U(1) and taking the minimal field content. We consider only discrete symmetries which act identically on all three generations. There are bound to be interesting models where this is not the case, but we wish to keep our analysis tractable and simple. There are seven varieties of irreducible representation (q, u^c, d^c, ℓ, e^c, h_1, and h_2) and 4 constraints: the operators qu^ch_1, qd^ch_2, ℓe^ch_2 and h_1h_2 must be allowed. This allows for a $Z_{N_B} \times Z_{N_L} \times Z_{N_R}$ discrete symmetry with transformation properties

$$q(\alpha), \quad u^c(\alpha^*\beta), \quad d^c(\alpha^*\beta^*), \tag{6}$$

$$\ell(\gamma), \quad e^c(\gamma^*\beta^*),$$

$$h_1(\beta^*), \quad h_2(\beta).$$

Z_{N_B} has group element α, with $\alpha^{N_B} = 1$, and is a discrete subgroup of B. Z_{N_L} has group element γ, with $\gamma^{N_L} = 1$, and is a discrete subgroup of L. Z_{N_R} has group element β, with $\beta^{N_R} = 1$, and is a discrete subgroup of T_{3R}. In the rest of this talk I will consider the case of just a single Z_N. I take $N_B = N_L = N_R = N$ and the Z_N is taken as the product of the above three Z_N.

To avoid renormalizable L violation

$$\beta^* \gamma \neq 1, \qquad (7)$$

and to avoid renormalizable B violation

$$\beta \alpha^3 \neq 1. \qquad (8)$$

At the outset it is important to point out that the electroweak breaking vevs, $\langle h_1 \rangle$ and $\langle h_2 \rangle$, do not lead to a problem with domain walls, even if $\beta \neq 1$. This is because the vacua labelled by $1, \beta, \beta^2, \ldots$ are related by a hypercharge gauge transformation: the wall between domains does not contain any energy.

The operators of dimension 5 allowed by (7) and (8) are:

$$[qqq\ell, u^c u^c d^c e^c]_F \sim \alpha^3 \gamma, \qquad (9a)$$

$$[qu^c e^c \ell]_F, \quad [h_1^+ \ell e^c]_D \sim 1, \qquad (9b)$$

$$[\ell\ell h_1 h_1]_F \sim \gamma^2 \beta^{*2}. \qquad (9c)$$

Those of (9b) cannot be forbidden. However they conserve B and L and therefore M_2 and are harmless. The most important point is that all the M_2 violating operators at dimension 5 are ruled out by (7) and (8). These include $[qqqh_2, qu^c e^c h_2]_F$ and $[d^{c+}qq, d^{c+}u^c e^c, e^{c+}h_1 h_1, h_1^+ h_2 e^c]_D$. At this order the LSP is still stable. However, the physics of this model can differ from the Z_2 case. If

$$\alpha^3 \gamma \neq 1, \qquad (10)$$

then B is conserved to this order, and the scale of the physics responsible for these operators can be quite low. This allows the neutrino masses of (9c) to be large enough to be interesting. This is unlike the Z_2 case where (9a) cannot be forbidden.

I now introduce a simple way of treating operators of high dimension which break B, L, or R_2. This does not require the additional freedom of β, which I now take to be unity. An operator of baryon number (b, ℓ) can be forbidden by

$$(\alpha^3)^b (\gamma)^\ell \neq 1. \tag{11}$$

An operator which allows proton decay has $b + \ell$ even, while one which allows LSP decay has $b + \ell$ odd. Conditions (7) and (8) are now

$$\gamma \neq 1, \quad \alpha^3 \neq 1, \tag{12}$$

which forbid $(0,1)$ and $(1,0)$ operators. The operators $(2,0)$, $(1,1)$, and $(0,2)$ conserve R parity and in order to discuss LSP decay we must examine $(3,0)$, $(2,1)$, $(1,2)$ and $(0,3)$. Of these the lowest dimension operators are:

$$(0,3) : [\ell^3 h_1^3]_F \tag{13a}$$

$$(1,2) : [qqq\ell\ell h_1]_F, [d^c d^c d^c \ell\ell]_F. \tag{13b}$$

In the former case $\tilde{\gamma} \to \nu\nu\nu$, which is an invisible decay, while in the latter case the $\tilde{\gamma}$ decays visibly to quarks and leptons.

Notice that proton decay can occur only via operators which are $b + \ell$ even, hence it is the R_2 conserving operators which must be forbidden by the discrete symmetries. The operators (13) do not allow proton decay, neutron oscillations or neutrino masses, so that the mass scale M responsible for these operators could be quite low.

The LSP decay rate is $\Gamma_{LSP} \sim m^5/M^4$ where m is the LSP mass:

$$\tau_{LSP} \sim 10^{-18} \sec \cdot \left(\frac{M}{10^4 \text{GeV}}\right)^4 \left(\frac{100\text{GeV}}{m}\right)^5 \tag{14}$$

Decay on a cosmological time scale would require a scale $M \leq 10^{11}m$. If the LSP is the photino with mass less than 10 GeV, further astrophysical constraints require $\tau < 10^5$ sec.[12], requiring a scale $M \lesssim 10^8 m$. If $M \lesssim 100$ TeV there will be no missing energy signatures in particle physical experiments even if the LSP is neutral.

I have shown that if a supersymmetric SU(3) $\times$ SU(2) $\times$ U(1) $\times$ Z_N model persists up to an energy scale above $10^{11}m$, without the addition of any extra fields, then the higher dimension operators do not give any significant LSP decay.

This may reinforce one's belief in a stable LSP. However, first it is necessary to examine the effects of extending the theory at the TeV scale.

A model in which baryon and lepton number violation first appear via a $(1,2)$ operator is usually considered to be artificial. In models where discrete symmetries Z_N arise naturally, such as in those inspired by superstrings, this is to be expected. Condition (12) implies that $\alpha, \gamma \neq 1$. Suppose $N = 5$ and $\alpha = \gamma = a, a^5 = 1$. The first baryon and lepton number violation is then $(13b)$ with $\alpha^3\gamma^2 = 1$. For $N = 6$ and $\alpha = \gamma = a, a^6 = 1$, two types of operators occur at dimension 8:

$$(2,0) : [qqqqd^{c^+}d^{c^+}]_D \qquad (15)$$

$$(1,3) : [u^c u^c u^c e^c \ell^+ \ell^+]_D.$$

The first allows neutron oscillation, while the second allows proton decay to three leptons. Our inability to observe B and L violating processes is often attributed to M being enormous; perhaps it is because N is not small.

Next we consider adding extra fields to the minimal set. There are a great many possibilities, and a completely general analysis is not possible. We show how the previous results are modified in a few particular cases. The power of the method should be clear.

Suppose the extra fields do not couple to quarks, leptons or Higgs. For a collection of Majorana fields X there will be mass terms $[XX]_F$, while for a Dirac pair the mass terms are $[XX^c]_F$. In either case, there is a parity, X-parity, which ensures that the lightest X particle is stable.

More interesting is the case when X couples to quarks and leptons. In many situations X can be assigned a lepton and baryon number such that B and L are conserved at dimension four. Higher dimension operators carrying (b, ℓ) allowed by Z_N can then be listed as before.

Consider the case of a quark which has charge -1/3 but is SU(2) neutral: $D(3, 1, -1/3)$ and $D^c(\bar{3}, 1, 1/3)$. If D^c has the same Z_N quantum number as d^c, then we have $[DD^c + qD^ch_2]_F$. To forbid renormalizable B and L violation: $\gamma \neq 1, \alpha^3 \neq 1$. This automatically forbids the new dangerous operators $[qqD + Du^ce^c]_F$. As before the $(0,2)$, $(1,1)$, and (2.0) operators conserve M_2 and so the LSP will decay at quite high dimension. In addition to $(13b)$ there are dimension 7 $(1,2)$ operators involving the exotic quark: $[DDD\ell^+\ell^+]_D$.

Another possibility is to introduce a pair of color neutral weak doublets: $L(1,2,-1/2)$ and $L^c(1,2,1/2)$. If they couple with $[\ell e^c L]_F$ then they are just like a pair of Higgs doublets. On the other hand if they couple with $[L e^c h_2]_F$, they are exotic leptons with Z_N quantum numbers $L(\gamma)$, $L^c(\gamma^*)$. There will be new operators involving L^c, but the counting is as before.

Another possibility is that D and D^c are leptoquarks rather than exotic quarks. This occurs if the renormalizable interactions are

$$[DD^c + Du^c e^c]_F. \tag{16}$$

In this case the Z_N transformations are $D(\alpha\gamma)$ and $D^c(\alpha^*\gamma^*)$ showing that they carry unit quark number (α) and unit lepton number (γ). $\gamma \neq 1$ rules out the usual $(0,1)$ operator together with $[qD^c h_2]_F$, while $\alpha^3 \neq 1$ rules out the usual $(1,0)$ operators. However there are other dimension 4 operators

$$[qqD, u^c d^c D^c]_F \sim (1,1) \qquad \gamma\alpha^3 \neq 1$$

$$[u^c D^c D^c]_F \sim (1,2) \qquad \gamma^2\alpha^3 \neq 1.$$

Thus the operators $(13b)$ are immediately ruled out in this model, and the LSP can first decay via $(0,3)$ operators.

An alternative possibility is to take $\gamma = 1$, $\alpha^3 \neq 1$, with $D(\alpha)$. This allows:

$$[DD^c, Du^c e^c, qD^c h_2, \ell h_1, \ell\ell e^c, qd^c \ell, qD^c \ell, \ell\ell h_1 h_1]_F. \tag{17}$$

All $(0,\ell)$ operators are allowed, while (b,ℓ) may easily be forbidden to quite high b (4 for $N = 4$, 5 for $N = 5$). The renormalizable lepton number violation leads to $\mu \to e\gamma$, neutrino masses, etc. but gives only fairly mild constraints on the coefficients of the $(0,1)$ operators. Of course the $(0,\ell)$ terms could also be allowed in the minimal supersymmetric model.[2] The LSP lifetime depends greatly on what the LSP is. The photino decays at one loop and would probably travel a measurable distance. On the other hand, if the LSP was a slepton it would decay to two leptons very rapidly.

All the examples given so far for X could occur in superstring inspired models. Each X had quantum numbers of a member of a 27 of E_6. However, the real superstring motivation for this work is that discrete symmetries Z_N, $N > 2$, occur very readily in these models.[9-11]

As a last example consider X to have charge 2: $X(1,1,2)$ and $X^c(1,1,-2)$. If $\gamma^2\alpha^3 = 1$, a $(1,2)$ operator can occur at dimension 5:

$$[XX^c, e^c e^c X^c, u^c u^c u^c X]_F. \tag{18}$$

The photino can decay into 3 up quarks and two electrons with a rate $\Gamma_{LSP} \sim m^3/M^2$, which can be very rapid.

There has been a theoretical bias that the LSP is stable. This has led to many supersymmetry searches based on missing energy, and has led to considerable work on the LSP as dark matter. It is certainly true that a simple way of building acceptable supersymmetric models is to make the LSP stable. However, I have argued that low energy supersymmetric theories with a discrete Z_N symmetry, $N > 2$, are also perfectly acceptable. These theories naturally account for our inability to uncover B and L violation. Indeed they often predict that the proton is stable, or that it has very unusual decay modes. A wide variety of LSP lifetimes is possible. Experimental searches for supersymmetry should bear in mind the new possibilities: that the LSP may decay before travelling a measurable distance, that it may travel an observable distance before decay, or that it may escape the apparatus having left a track. The last possibility appears very likely, and would correspond to the LSP being a long-lived charged slepton or squark (R meson).

R parity breaking remains a very important question for supersymmetry. I have shown that R parity may just be a low energy accident due to the presence of a Z_N symmetry.

Acknowledgments

I thank the organizers of the *Quarks and Galaxies* workshop held at LBL for a very stimulating meeting. The idea of generalizing R parity to Z_N was invented at this workshop in discussions with Graham Ross.

This work was supported in part by NSF Grant PHY82-15249, by a Sloan Fellowship and also in part by the Director, Office of Energy Research, Office of High Energy and Nuclear Physics, Division of High Energy Physics of the U.S. Department of Energy under Contract DE-AC03-76SF00098.

References

1. That R parity is sufficient to yield a stable particle was realized by G. Farrar and P. Fayet, Phys. Lett. 76B (1978) 575. The necessity for a matter parity in modern supersymmetric models was realized by S. Dimopoulos and H. Georgi, Nucl. Phys. B193 (1981) 150.

2. L.J. Hall and M. Suzuki, Nucl. Phys. B231 (1984) 419.

3. I-Hsiu Lee, Nucl. Phys. B246 (1985) 120.

4. F. Zwirner, Phys. Lett. 132B (1983) 103.

5. R. Barbieri and A. Masiero, Nucl. Phys. B267 (1986) 679.

6. G.G. Ross and J.W.F. Valle, Phys. Lett. 151B (1985) 375.

7. J. Ellis, G. Gelmini, C. Jarlskog, G.G. Ross, and J.W.F. Valle, Phys. Lett. 150B (1985) 142.

8. I thank B. Grinstein for pointing this out to me, and G. Gelmini for conversations.

9. L.J. Hall and G.G. Ross, preprint in preparation.

10. E. Witten, Nucl. Phys. B258 (1985) 75.

11. B. Greene, K. Kirklin, P. Miron, and G. Ross, Oxford preprint, May 1986.

12. A. Bouquet and P. Salati, preprint (1986) LAPP-TH-157.

COSMIC STRING SEARCHES

Craig J. Hogan*

Steward Observatory, University of Arizona

Tucson AZ 85721, USA

ABSTRACT

I discuss observational strategies for finding effects associated with the gravitational lensing of distant objects by strings. In particular, the requirements of a survey to find chains of galaxy image pairs or single galaxies with sharp edges are studied in some detail, and a proposed search program at Steward Observatory is described.

Cosmic strings have emerged as possibly an important constituent of our universe, and may be the root cause of all galaxy formation and clustering.[1-7] Independent observational tests of the cosmic string scenario, such as discontinuities in microwave background temperature[8] or pulsar timing noise induced by strings through gravitational radiation[7], are precise enough to give it a level of respectability (as determined by falsifiability) on a par with other theories of galaxy formation. But the string scenario can be verified in principle at a much higher level of certainty by actually finding an actual string. Here I explore the possibility of satisfying a skeptical astronomer who says "show me one." For a more detailed account of the string scenario in general and string-induced galaxy formation in particular, see R. Brandenberger's lecture in this volume or the above references. I will adopt the standard notation, where the mass per unit length of string is denoted by μ, and $\mu_{-6} \equiv G\mu/10^{-6}c^2$ is a convenient parameterization; for galaxy formation $\mu_{-6} \simeq 2$ is the currently preferred value[3], but this depends on uncertain factors such as the composition of the dark matter.

*Alfred P. Sloan Research Fellow

Optical searches can reveal a number of unique string-like effects. (1) Double images of QSO's can be formed by strings, with a string passing between the images.[9,10,11] This in itself is a stringy feature, as only a singular lens like a string can induce an even total number of images. In addition, velocities viewed on opposite sides of the string are shifted[8,11] by $\simeq 8\pi G\mu \simeq 8\mu_{-6}$ km/sec. Thus common Ly$-\alpha$ absorption lines in the two images should be systematically shifted in redshift by observable amounts, in some cases comparable to their width. For lines observed in the two images of QSO 2345+007A,B by Foltz et al. [12] the rms measuring error in comparing redshifts in the two images is about 25 km s^{-1}, so a test for a systematic shift should be possible with better data now being obtained. (2) Multiple (even) numbers of QSO images can be formed around compact string loops, whose rapid oscillatory behavior leads to rapid time variability. Rapid appearance and disappearance of images is also possible, and various features of the image appearance and brightness variations have a unique signature predicted from catastrophy theory.[13] (3) Long straight strings passing in front of a distant field of galaxies create a chain of galaxy pairs (with one image on each side of the string) whose angular separations are similar in magnitude $(\delta\phi \approx 4\pi G\mu)$ and whose position angles (as defined by the separation of the paired images) are nearly identical, although they are not in general perpendicular to the string.[11,14] (4) If the string itself actually passes in front of an extended resolved object of angular extent less than $4\pi G\mu$, it can produce an image with a sharp edge.[15]

The phenomenon of QSO lensing by strings has been extensively discussed elsewhere[9-11,13] and the search for stringlike effects is included in various programs studying and searching for lensed QSO's in general.[16] On the other hand the possibility of searching for lensing of galaxies by strings has been relatively neglected. Therefore, in this talk I will concentrate on optical search strategies for the effects (3) and (4) just listed. How they arise is illustrated in figure 1. Figure 1a shows a toy geometry with two distant galaxies, A and B, and a row of numbers floating in space. Light rays are shown bending around the string with angle $\theta_\mu \equiv 4\pi G\mu = 2.6\mu_{-6}$ arc sec, which is independent of impact parameter. Figure 1b depicts the appearance of the sky. The strip between numbers 3 and 4 can be seen around either side of the string. Galaxies such as A lying within this strip have duplicated images. Galaxies such as B which straddle the edge of this strip are duplicated also but one image is only a duplicate of the part which lies inside the strip, and hence has a sharp edge. For a string with

118

typical velocity $\simeq c/2$, the position angles separating the images in each pair
would not be perpendicular to the projected location of the string, although for
a nearly straight string these position angles would be nearly parallel.[11]

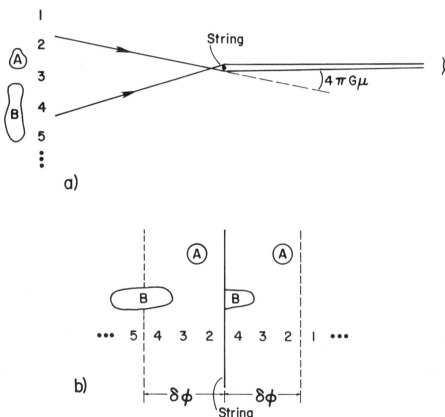

a)

b)

Consider the requirements for a search for a strip of duplicated galaxies.
One must first have enough galaxies in one's survey to have a good chance of
several of them landing in the narrow strip where they will be lensed. The
fraction of the sky covered by string strips varies only logarithmically with the
area of sky surveyed, and is approximately[13,14] $P \simeq 60G\mu$. Therefore, almost
independent of the survey area one needs $3/60G\mu \simeq 5\times10^4\mu_{-6}^{-1}$ galaxies in order
to have three of them land within a string strip. (One stands the best chance of
these galaxies belonging to the *same* string segment if the survey geometry is
itself striplike, which also facilitates follow-up searches as candidates can easily
be verified by deep searches on either side of the search strip.) Thus in figure

2 solid horizontal ("3 pairs") lines indicate the minimum total galaxy number for this type of search, for several values of μ_{-6}.

Galaxies must not only lie within the correct strip, they must also typically lie behind the strings. This will be true for a significant fraction of galaxies only if the survey depth extends out to of order the Hubble distance. For a survey limit of J magnitude 24, the typical sample depth is about 2000 h^{-1} Mpc[17] and the typical galaxy density is about 10^4 per square degree.[18,19] This is shown as another lower limit in fig. 2, a line of constant surface density for 24 mag galaxies.

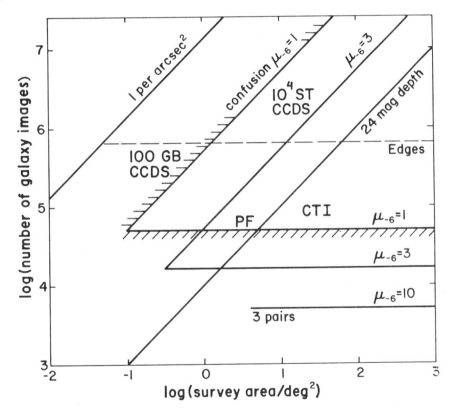

Finally, there is the possibility of confusion if the galaxy density in the survey is too high. It will be difficult to select the 3 string-induced pairs out of the haystack of random galaxy pairs (even with the alignment constraint) unless the mean separation of galaxies in the survey is much larger than the string-induced image separation. The confusion limit is shown in fig. 2 for two values of μ_{-6}, assuming the mean projected density must be less than one per $(8\pi G\mu)^2$ sr. Both the number and confusion constraints are somewhat harder to satisfy than these estimates would indicate because of galaxy autocorrelations, which reduces the number of effectively independently-sampled images and adds to the likelihood of close pairs without strings. Nevertheless, these numbers are useful as rough guides, as correlations are small in very deep samples.

It should be mentioned that *any* field of the sky larger than a certain minimum size is likely to have at least one string loop in the field of view at $z \lesssim 1$. The total angular length of string ℓ at $z \lesssim 1$ in any survey field is proportional to the survey area A. Using our previous estimate, we have (still ignoring the log A dependence of P^{13})

$$(A/4\pi \ sr) \cdot 60 G\mu = 4\pi G\mu(\ell/rad)$$

$$(\ell/rad) \simeq 0.38(A/sr)$$

This length is distributed evenly in equal octaves of ℓ which exceed the gravitational-wave-decay limit. Loops whose gravitational-decay timescale is comparable to H^{-1}, placed at a distance $ct_o/2$, have an angular circumference[13] $\simeq 4\pi G\mu \simeq 2.6\mu_{-6}$ arc sec, and hence have a mean density of about one per $4\pi G\mu/.38 \simeq 3 \times 10^{-5}\mu_{-6}$ sr $\simeq 400$ (arc min)$^2\mu_{-6}$. Any field larger than $20\mu_{-6}^{1/2}$ arc min on a side is thus almost certain to contain a loop within about 2000 h^{-1} Mpc. A field 1000 times bigger is almost certain to contain a loop 10 times closer; thus, a survey of 100 deg^2 need only have a depth of 200 h^{-1} Mpc, or about 19th mag, to contain a loop. The angular size of image separations around a straight segment of loop does not depend strongly on the survey depth[11]:

$$\delta\phi = d_{gs}(d_{gs} + d_{os})^{-1}8\pi G\mu\sin\theta$$

where d_{gs} and d_{os} are the galaxy-string and observer-string separations and θ is the inclination of the string to the line of sight. Since the width of the lensing strip on the sky is independent of survey depth, one needs more galaxies in a shallow survey because the total string length is proportional to the survey volume; even if the field is guaranteed to contain a string, one must still find enough background objects to see where it is. Nevertheless, these considerations indicate the 24 mag criterion is not absolute, and somewhat shallower surveys can compensate by covering a wider area.

A square field $\theta_f \gtrsim 20$ arc min on a side in general contains a spectrum of loops, the largest of which however is still usually smaller $(\simeq 0.4\theta_f^2$ rad in circumference) than the field itself. A long strip of sky on the other hand in general intersects many loops larger than its thin dimension even though the length of string per area is the same as the square field. Exactly such a strip of deep digitized images will be provided by Steward Observatory's CCD transit instrument (CTI)[20] whose projected performance is contrasted in fig. 2 with digitized deep prime-focus photographic surveys[18] (PF) and other possible surveys using multiple CCD exposures on large ground-based telescope (100 GB CCD's)[19] or several years of serendipitous WFPC images on the space telescope (10^4 ST CCD's).[15] Note that the CTI search is not working to the limits of ground-based detectors so follow-up images can be taken to verify candidates found in the survey by looking deeper and also by looking for additional pairs produced by the same string to either side of the strip. I have not included here the constraint arising from ground-based seeing; this will make it impossible to detect image pairs closer than $\sim$1 arc sec, limiting the search to $\mu_{-6} \gtrsim 0.3$.

For finding edges on galaxies at $z \gtrsim 1$, ST is necessary just to achieve the necessary resolution. Here the number density of galaxies n required depends on the size of galaxies θ_g, not on $G\mu$ (the angular length of string ℓ is independent of $G\mu$); to find an edge, we require the two-dimensional projected "optical depth" $n\,\theta_g\,\ell > 1$, and using the above formula for $A(\ell)$,

$$A \gtrsim 80 \deg^2 (n/10^4 \deg^{-2})^{-1} (\theta_g/\text{arc sec})^{-1}$$

This translates again into a minimum required number of galaxies, now almost a million ("EDGES" in fig. 2). But as Paczynski[15] points out, even this is within the capability of ST. Even with ST, another resolution problem arises if $4\pi G\mu \lesssim 1$ arc sec; sharp edges will no longer be conspicuous because the fuzzy image of a galaxy will fill the entire strip and the clipped image will merge with the unclipped one. Thus in practice this effect is also seen only for $\mu_{-6} \gtrsim 0.3$.

To my knowledge, of all the possibilities just outlined the only search seriously being planned for the near future is the one we are planning at Steward using CTI. The collection of data by this instrument is eventually to be entirely automated, so the search for cosmic strings, like CTI searches for many other interesting phenomena, boils down to running various algorithms on an existing data base, with follow-up on ordinary telescopes. It is possible that CTI will collect enough images of distant galaxies by coadding data for a year or so that even with a non-detection of correlated image pairs to a specified level we will be able to put useful constraints on the string scenario. These constraints may be expected to be complementary to those already existing from microwave and gravitational-wave techniques.

This work was supported by the Alfred P. Sloan Foundation and by NASA grant NAGW-763 at the University of Arizona.

REFERENCES

1. Zeldovich, Ya.B., Mon. Not. R. astr. Soc. 192, 663 (1980).

2. Vilenkin, A., Phys. Rev. Lett. 46, 1169 (1981); 46, 1496(E) (1981).

3. Turok, N. and Brandenberger, R.H., Phys. Rev. D. 33, 2175 (1986).

4. Turok, N., Phys. Rev. Lett. 55, 1801 (1985).

5. Vilenkin, A., Phys. Rep. 121, 263 (1985).

6. Vilenkin, A., "String Review", in Inner Space/Outer Space, Kolb, E.W. et al. , eds. (Chicago: Univeristy of Chicago Press, 1986), p. 269.

7. Hogan, C.J. and Rees, M.J., Nature 311, 109 (1984).

8. Kaiser, N. and Stebbins, A., Nature 310, 391 (1984).

9. Vilenkin, A., Phys. Rev. D. 24, 2082 (1981).

10. Gott, J.R., Astrophys. J. 288, 422 (1985).

11. Vilenkin, A., Nature 322, 613 (1986).

12. Foltz, C.B., Weymann, R.J., Röser, H.-J., and Chaffee, F.H., Astrophys. J. 281, L1 (1984).

13. Hogan, C. and Narayan, R., Mon. Not. Roy. astr. Soc. 211, 575 (1984).

14. Vilenkin, A., Astrophys. J. 282, L51 (1984).

15. Paczynski, B., Nature 319, 567 (1986).

16. Turner, E.L., "Gravitational lenses and dark matter: observations" (preprint, Princeton University Observatory), 1986.

17. Peebles, P.J.E., The Large Scale Structure of the Universe (Princeton, 1980).

18. Ellis, R. in The Origin and Evolution of Galaxies, ed. B. Jones and J. Jones (Reidel, 1981), p. 255; Tyson, J.A. and Jarvis, J.F., Astrophys. J. 230, L153 (1979).

19. Tyson, J.A., Seitzer, P., Weymann, R.J., and Foltz, C., Astron. J. 91, 1274 (1986); Seitzer, P., Tyson, J.A., and Butcher, H., Astron. J., in press (1986).

20. McGraw, J.T., Stockman, H.S., Angel, J.R.P., Epps, H. and Williams, J.T., Opt. Eng. 23, 210 (1984).

AN ATTEMPT TO RELATE
THE LINEAR EVOLUTION OF PRIMORDIAL PERTURBATIONS
TO THE NON-LINEAR STRUCTURE OF THE UNIVERSE

F. Occhionero
Istituto Astronomico, Università di Roma I
via G.M.Lancisi, 29 - 00161 Roma,ITALY

R. Scaramella
SISSA
Strada Costiera 11, 34014 Trieste, ITALY

ABSTRACT
Following a suggestion by Schaeffer and Silk,
we attempt to calculate the present distribution
of galaxies from the linear growth of primordial
adiabatic fluctuations. The latter are studied
in Universe models dominated by dark matter, in
either one of its three popular varieties - hot,
warm, and cold - or in some mixture thereof
(hybrid models).
    The process of hierachical clustering seems
well accounted for at the stage of non-linearity;
in principle, the comparison of these results
with the Schechter counts should be possible
under the assumption that galaxies are not
destroyed once created. On the contrary, this
method seriously underestimates the reality, if
fragmentation is an important process of galaxy
formation.

1.   INTRODUCTION
    The purpose of the present communication is twofold:

1) to report and improve on ongoing numerical work[1] on the evaluation

of linear adiabatic modes in cosmological models dominated by two

forms of dark matter (DM), of which one is free streaming and the

other is not (hybrid);

2) to experiment with a specific formalism[2,3] proposed to infer

from our knowledge on the linear growth the evolution of the non-

linear structure.

We are assuming that DM is an essential ingredient of cosmological models and, following the established conventions[4], we will term
it hot (H), warm (W), and cold (C) : of these the first two are free
streaming and cluster respectively above the scale of the super-
cluster ($\sim 10^{15}$ M_Θ) and of the large galaxy ($\sim 10^{12}$ M_Θ), while the
third clusters on all scales by definition. For the present astronom-
ical applications, we disregard the explicit microscopical nature of
DM; we recall[5] that even WDM has its candidates (a 1 keV gravitino,
a 1 keV right-handed neutrino, a 1 keV axino), although of course
massive neutrinos (on the side of HDM) and heavy particles (such as
1 GeV photinos and gravitinos) and the ultralight axions (on the
side of CDM) are much more popular with particle physicists.

On the basis of inflation[6,7,8] we are here accepting the
theoretical prejudice that

$$\Omega_o = 1,$$

where Ω_o is the usual ratio of the present to the critical density.
This implies immediately that DM is dominant since baryonic matter
is scarce ($\Omega_{oB} \ll 1$) both observationally and theoretically[9]. In
fact, in the sequel we will ignore completely the dynamical effects
of the latter. Our basic parameter is the fraction of the CDM
density to the total density,

$$\eta = \Omega_{CDM} \quad ;$$

obviously HDM or WDM accounts for the remaining $(1-\eta)$ fraction of
the total density.

In the case of HDM, we assume that only one of the three
neutrino flavors is massive; its mass is then of the order of
$90(1-\eta)$ eV. This neutrino is sufficiently heavy to be non-relativistic
at equivalence.

In the case of WDM, we assume a WIMP with a mass of the order
of 750 $(1-\eta)$ eV and the existence of three massless neutrinos.

Furthermore, for the sake of simplicity, we take $h=H_o/100$ km/sec /Mpc to be unity.

2. FREE STREAMING

We have integrated numerically the differential equations for the linear adiabatic perturbations written in the familiar synchronous gauge [10,11]. Baryonic matter and radiation were treated as a single relativistic fluid and were neglected altogether after matter-radiation decoupling; however baryons are assumed to fall thereafter into the potential traps of CDM.

Concerning the perturbed part of the distribution function of the free streaming component, as in previous work [12], we have chosen

i) to sample momentum space at a number N of values suitable for the necessary Gauss-Laguerre integrations, and

ii) to expand the angular dependence in Legendre polynomials, truncating the expansion at $l=L$.

In order to save computer time, L was chosen for each wavenumber $(k \propto M^{-1/3})$ in such a way to attain an accuracy better than 1% (the accuracy being defined as the difference between the results on the dominant component of the L-th and the (L-2)-th approximation). Clearly L increases quickly below the free streaming mass. We have compared our results with the analytical fit given by Bardeen et al.[17] in the case of pure HDM ($\eta=0$ in our notation) and found the agreement to be excellent; in fact we believe our numerical results to be accurate down to 10^{13} M_Θ in the case of pure HDM and down to 10^{10} M_Θ in the case of pure WDM. On the other hand, for $\eta=1$, our computer code reproduces correctly the familiar pure CDM model[13]. We have also checked our results against those of a preexisting code[14] built along the lines suggested by Peebles[15] and found again excellent agreement.

The interesting physics here is in the evolution of the density fluctuation of the CDM component: as each perturbation enters its horizon, the growth of $(\delta\rho/\rho)_{CDM}$ undergoes a stagnation which is more pronounced the larger is $(1-\eta)$, i.e. the more important is the free streaming component. This stagnation ends and normal growth resumes in the matter dominated era only when even the free streaming component starts to fall into the condensation (its decreasing Jeans mass, $\propto(1+z)^{3/2}$, being eventually overtaken by the perturbation mass). This occurs progressively later, the smaller is the mass; thus the transmission factor $T(k)$ falls faster than k^{-2} for large k. This is seen in Fig.1 where $T(k)$ is shown vs. k in Mpc^{-1}; in the left panel we have the HDM models, in the right panel the WDM models; in either case the broken lines refer to $\eta=0$, full lines to $\eta=50\%$. In addition the pure CDM model is given in both panels for reference with dotted lines.

3.   CONDENSATIONS

We calculate the r.m.s. density fluctuation through the well known formula

$$\Sigma^2(M,z) = A \int d^3k \; W(kR) \; P(k,z)$$

where $P(k,z)$ is the power spectrum resulting at any redshift z from our numerical integrations and from the initial power spectrum with spectral index $n=1$, and where the window function

$$W(x) = \{3(\sin x - x \cos x)/x^3\}^2$$

is appropriate to a sharp sphere of radius R containing the mass M. Finally the normalization constant A is chosen in all cases so to normalize the variance at $8 \; h^{-1}$ Mpc in agreement with the galaxy counts[16] (no bias).

In Fig.2 we show the variance Σ vs. mass in solar units at three different redshifts, $1+z=1,4,10$ from top to bottom; left and right panels are organized as in Fig.1.

In the left panel the alternative top-down and bottom-up scenarios (for pure HDM and pure CDM) arise from the different slopes of the broken and of the dotted curves. The hybrid model (full lines) appears at this stage more like pure HDM, than pure CDM, but it will show a behavior quite distinct from either one in the formation of condensations. In the right panel, WDM appears characterized by the known [19] fact that it gives rise to structure on all scales below 10^{12} $M_\odot$ at a redshift of 10 (when the variance attains unity regardless of the chemical composition).

Following Schaeffer and Silk [2,3] and Press and Schechter [18], we now assume a Gaussian probability for any given density contrast δ_M at any redshift z

$$\Pi(\delta_M ,z) = \frac{1}{\sqrt{2\pi}} \frac{1}{\Sigma(M,z)} \exp \{- (\delta_M^2/2\Sigma^2)\} \quad .$$

Then the fraction of the cosmic matter that has condensed (i.e. gone non-linear) at redshift z is given by

$$F(M,z) = \int_1^\infty d(\delta_M)\Pi(\delta_M , z)$$

and the distribution of mass in condensed objects vs. mass is given by

$$\Phi(M,z) = - M (dF/dM) \quad .$$

As the authors themselves point out, the above equation is just a working hypothesis, attractive for its simplicity: on one side, it is simpler than counting the density maxima which we would like to associate with galaxies, but on the other side it does not account for substructure, the so-called "cloud-in-cloud" problem [18].

We present our preliminary results on the distribution Φ for the three redshifts 1+z=1,4,10 in Fig.3. The interpretation of the latter is less straighforward; let us begin from the pure CDM models shown in both panels by the dotted lines. As expected in the general bottom-up scenario, it is seen here that power appears early $(1+z=10)$: with time it both moves to larger scales and disappears from the galactic one (in other words the curves shift to the right). This should not be read as saying that galaxies actually disappear - in fact, only occasionally will they be destroyed in collisions and mergings -, but that they become part of larger systems.

Let us now consider in alternative the free streaming models beginning with the HDM cases of the left panel. For the pure model (broken lines), as one expects from the top-down scenario which applies here regardless of our normalization, power shows up significantly only at the present in a peak at the supercluster scale, but is totally absent earlier on; in fact, on the figure we have a graph for $1+z=4$, but not for $1+z=10$. We may say that in this case the distribution curves shift upward with time; anyhow we have far fewer galactic objects than is required[2,3]. The hybrid case, $\eta=50\%$ (full lines), combines interestingly the two previous features in that the distribution curves first move up, then to the right: structure is present on the galactic scale already at $1+z=4$ (while there is practically none at $1+z=10$) and then shifts to larger masses via clustering. The numbers being intermediate between the pure HDM and the pure CDM cases, this model is likely to fit better the Schechter counts[3].

The WDM models are shown in the right panel: the pure model $(\eta=0)$ differs markedly from the pure CDM model on the small scales, where the very sharp slopes imply that there are fewer objects than required[3]. At late times, hierarchical clustering goes on as in the pure CDM model. The WDM hybrid $(\eta=50\%)$ contains many more low mass objects at $1+z=10$ than the HDM hybrid.

4. ACKNOWLEDGMENTS

F.O. is grateful for hospitality to the Dept. of Astronomy of UCB where this work was begun, for discussions to J.R.Bond, M.Davis, J.Silk and to the members of the LBL Workshop, for financial support to the Italian Ministry of Public Education, for travel support to the Italian Foreign Ministry.

5. REFERENCES

1. S. Achilli, F. Occhionero, R. Scaramella, Ap.J. 299,577 (1985)

2. R. Schaeffer, J. Silk, Ap.J. 292, 319 (1985)

3. R. Schaeffer, J. Silk, Ap.J. in press (1986)

4. J.R. Bond, A. Szalay, Ap.J. 274, 443 (1983)

5. M.S. Turner, in Dark Matter in the Universe, eds. J. Knapp and J. Kormendy (Reidel, 1986)

6. A. Guth, Phys.Rev.D. 23, 347 (1981)

7. A. Linde, Phys.Lett.B 108, 289 (1982)

8. A. Albrecht, P. Steinhardt, Phys.Rev.Lett. 48, 1220 (1982)

9. J. Yang et al., Ap.J. 281, 493 (1984)

10. S. Weinberg, Gravitation and Cosmology (J. Wiley, 1972)

11. P. J. E. Peebles, The Large Scale Structure of the Universe (Princeton, 1980)

12. S. Bonometto et al.,Astr.Ap. 138, 477 (1984)

13. G. Blumenthal et al., Nature 311, 517 (1984)

14. R. Scaramella, Thesis, U.of Rome (1984)

15. P. J. E. Peebles, Ap.J. 258, 415 (1982)

16. P. J. E. Peebles, Ap.J.Lett. 263, L1 (1982)

17. J. Bardeen et al., Ap.J. 304, 15 (1986)

18. W. Press, P. Schechter, Ap.J. 187, 425 (1974)

19. J. R. Bond, A. S. Szalay, M. S. Turner, Phys. Rev.Lett. 48, 1636 (1982)

130

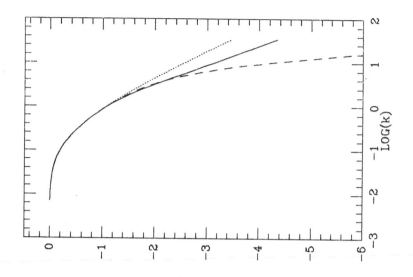

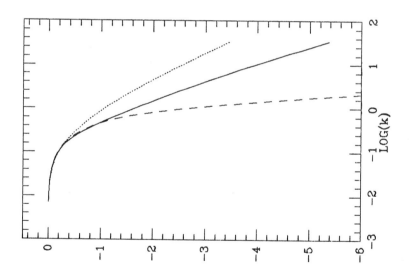

Fig. 1   Log-log plot of the transmission factor T(k) vs. wavenumber
in Mpc^{-1}for HDM cosmologies (left) and WDM cosmologies
(right). Dotted lines refer to the pure CDM model shown
as a reference, full lines to η=50%, broken lines to η=0.

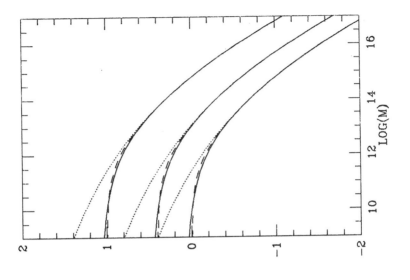

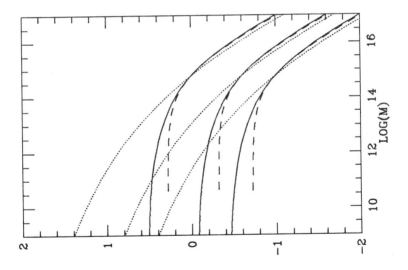

Fig. 2   Log-log plot of the variance Σ vs. mass in solar units at the three redshifts 1+z=1,4,10 from top to bottom; left and right panels are organized as in Fig. 1.

132

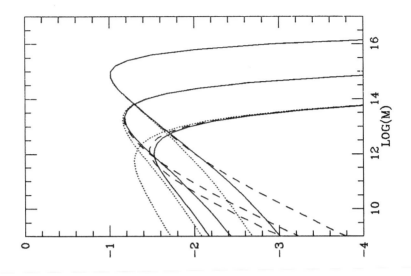

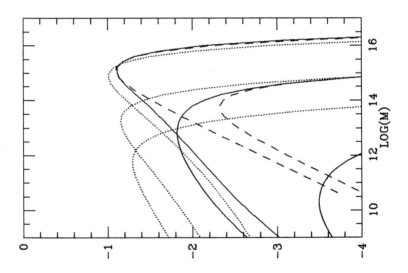

Fig. 3   Log-log plot of the distribution function Φ vs. mass in
          solar units for the three redshifts of Fig. 2. On the low
          mass side the total distribution has the sharpest slope
          for the pure free streaming models.

Superstring Candidates for Dark Matter

Keith A. Olive
School of Physics and Astronomy
University of Minnesota
Minneapolis, MN 55455

Abstract

I will discuss some of the possible candidates which naturally
arise in low energy superstring-inspired models. In addition, I
will also discuss the possibility of detecting cold dark matter
using under-ground proton-decay detectors searching for high
energy (E > 1 GeV) neutrinos from the annihilations of dark
matter inside the sun.

As one can see from these proceedings, one of the major topics in
cosmology and particle physics today is dark matter: its identity,
role and possible detection. My "official" topic for a contribution
to these proceedings is on the identity of dark matter candidates
found in low energy extrapolations of the superstring. Before I address this issue, however, I would like to discuss in some further
detail the possibility discussed by Mark Srednicki[1] of detecting
cold dark matter candidates by searching for high energy neutrinos
coming from dark matter annihilations inside the sun[2-5].

Basically, if we assume that the galactic halo is dominated by nonbaryonic cold dark matter at a density $n_\chi = (0.3 \text{ GeVcm}^{-3})m_\chi^{-1}$ in the solar neighborhood with an average velocity $v \approx 300 \text{ kms}^{-1}$, elastic scatterings off of protons in the sun will trap dark matter particles in the sun at a rate[6]

$$\Gamma_T \simeq 7 \times 10^{28} \text{ s}^{-1} \sigma_{p,36} m_\chi^{-1} \tag{1}$$

where the elastic cross-section for $\chi + p$ is $\sigma_p = (10^{-36} \text{cm}^2)$ $\sigma_{p,36}$. The sun will lose dark matter particles through either evaporation[7] or annihilation[8,2]. For $m_\chi > 6$ GeV, it was found[1] that annihilations dominate over evaporations. The only annihilation products capable of leaving the sun are of course neutrinos, which may be detectable[2-5]. For further details, the reader should see Refs. 1-5).

Of the "standard" dark matter candidates[3], Dirac and Majorana neutrinos, sneutrinos, photinos, higgsinos, etc., I will concentrate here only on the latter two supersymmetric choices. In general, one would not really expect massive neutrinos to be stable, i.e., there is no symmetry protecting their decay. In supersymmetric theories, R-parity (see below) guarantees the existence of one stable particle. Although the sneutrino[9] may be the lightest supersymmetric particle (LSP), it in general requires a fine tuning of mass parameters so that indeed the sneutrino is the LSP. Furthermore, terrestrial experiments[10] seem to rule out sneutrinos and Dirac neutrinos for masses in the range 16-1000 GeV. Thus the only particles I will consider here are photinos[11,12] and higgsinos[12].

In his contribution [1], Mark Srednicki discussed the general calculations for the fluxes of neutrino coming from the dark matter annihilations. Here I will restrict my attention to the prompt neutrinos [4,5] coming from $\chi\chi \to ff; f \to f' + \ell + \nu_\ell$ where $f = \tau$, c or b (and possibly t if $m_\chi > m_t$). It is straight forward to calculate the differential neutrino flux from the sun[5]

$$\frac{d\Phi}{dE_\nu} = \sum_f \frac{(1/2) \; \Gamma_T \; B_f}{4 \; \pi \; d^2} \; \frac{1}{\Gamma_f} \; \frac{d\Gamma_f}{dE_\nu} \tag{2}$$

where $\frac{1}{2}\Gamma_T$ is equal to the annihilation rate (in equilibrium), d is
1 A.U., and Γ_f and B_f are the decay rates and branching ratios for
$f \rightarrow f' + \ell + \nu_\ell$. The background that we must compare the spectrum (2)
to is the flux of atmospheric neutrinos[13] due to cosmic rays.

Because annihilations in the sun will produce equal fluxes of ν_e,
$\bar{\nu}_e$, ν_μ and $\bar{\nu}_\mu$, while the atmospheric background flux is larger for ν_μ +
$\bar{\nu}_\mu$ than for $\nu_e + \bar{\nu}_e$, it is advantageous (if possible) to compare the ν_e
+ $\bar{\nu}_e$ fluxes. In addition, the atmospheric flux is expected to be
isotropic at high energies so that to maximize the signal to background
ratio one should correlate observed events with the position of the sun.

As I stated earlier, the dark matter candidate I wish to consider
is a supersymmetric particle guaranteed to be stable because of an
unbroken R-parity. R-parity can be defined in terms of the quantum
numbers, spin, lepton number and baryon number

$$R = (-1)^{2S + L + 3B} \tag{3}$$

so that all "normal" particles, quarks, leptons and gauge bosons have
R = +1, while all "supersymmetric particles, squarks, sleptons and
gauginos have R = -1. Conservation of R-parity implies that the
lightest R = -1 state must be stable (i.e., the LSP). The most
plausible choice for the LSP is some linear combination of the R = -1
neutral fermions[12]

$$\chi = \alpha \tilde{W}^3 + \beta \tilde{B} + \gamma \tilde{H} + \delta \tilde{\tilde{H}} \tag{4}$$

where $\tilde{W}^3$ is the third component of the wino, $\tilde{B}$ is the bino and $\tilde{H}$ and $\tilde{\tilde{H}}$
are two higgsinos. In terms of $\tilde{W}^3$ and $\tilde{B}$ the photino is defined by

$$\tilde{\gamma} = \frac{g_1 \tilde{W}^3 + g_2 \tilde{B}}{\sqrt{g_2^2 + g_2^2}} \tag{5}$$

where g_1, g_2 are the U(1) and SU(2) gauge couplings respectively.

136

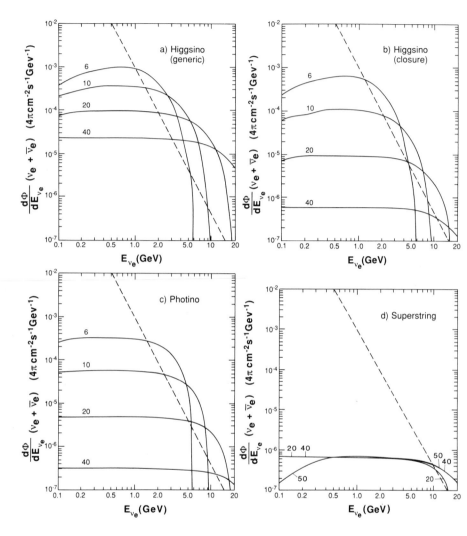

Figure 1: The differential flux of $\nu_e + \bar{\nu}_e$ from the sun due to the annihilations of cold dark matter as compared to the atmospheric background flux in the direction of the sun (dashed line) for
a) a generic higgsino, $\chi = \tilde{H}$ or $\tilde{\tilde{H}}$;
b) a symmetric higgsino $\chi = (v\tilde{\tilde{H}} + \bar{v}\tilde{\tilde{H}})/(\; v^2 + \bar{v}^{\;2})^{1/2}$;
c) a photino and
d) a superstring LSP.

Plotted in Figs. 1a, b, c are the differential neutrino fluxes for three choices of an LSP: a) $\chi = \tilde{H}$; b) $\chi = \tilde{S} \equiv (v\tilde{\bar{H}} + \bar{v}\tilde{H})/(v^2 + \bar{v}^2)^{1/2}$ where $v = <0|H|0>$ and $\bar{v} = <0|\bar{H}|0>$ are the vacuum expectation values of H and $\bar{H}$ respectively and c) $\chi = \tilde{\gamma}$. Plotted in the figure is the sum $d\Phi/dE_{\nu_e} + d\Phi/dE_{\bar{\nu}_e} = 2d\Phi/dE_{\nu_e}$. There is an equal contribution from the muon neutrino flux. Also included in the figures (dashed) is the differential flux of atmospheric neutrinos taken from ref. 13. Only the flux of $\nu_e + \bar{\nu}_e$ is plotted. Although the solar flux of $\nu_\mu + \bar{\nu}_\mu$ equals the flux of $\nu_e + \bar{\nu}_e$, the atmospheric background flux of $\nu_\mu + \bar{\nu}_\mu$ ranges from $\simeq$ 2-7 times larger than the $\nu_e + \bar{\nu}_e$ flux. The isotropic flux in ref. 13 has been multiplied by a factor $2\pi(1 - \cos 30^\circ) \simeq$ 0.8sr for the flux coming from a cone in the direction of the sun with a half opening angle of 30°, which is meant to represent the capabilities of proton-decay detectors to determine the directionality of neutrino induced events.

To compare the signal to background it is useful to define the quantity

$$R \equiv d\Phi/dE_{\nu_e}\big|_{\odot} / d\Phi/dE_{\nu_e}\big|_{ATM} \qquad (6)$$

as the ratio of the flux due to annihilation to the atmospheric background flux is the 30° cone in the direction of the sun. From Fig. 1a, we see that for $E_\nu > 1$ GeV, the solar flux of neutrinos exceeds the atmospheric background (in the direction of the sun) for $m_{\tilde{H}} = 6$ GeV. Even for $m_{\tilde{H}} = 40$ GeV, $R > 1$ when $E_\nu \geq 3$ GeV. At its maximum, $R \simeq 10$ for $m_{\tilde{H}} = 6$ GeV at $E_\nu \simeq 3$ GeV; $R \simeq 15$ for $m_{\tilde{H}} = 10$ GeV at $E_\nu \simeq 4$ GeV; $R \simeq 40$ for $m_{\tilde{H}} = 20$ GeV at $E_\nu \simeq 7$ GeV and $R \simeq 120$ for $m_{\tilde{H}} = 40$ GeV at $E_\nu = 20$ GeV.

For $\chi = \tilde{S}$, it is possible to arrange for a decoupling of the LSP higgsino from the Z° when the ratio of vev's $v/\bar{v} \simeq 1$. In Fig. 1b, we plot the neutrino fluxes resulting from such an LSP S, adjusting the ratio $v/\bar{v}$ so that $\Omega = 1$ (for $h_\circ = 1/2$). Thus for $m_{\tilde{S}} = 6$, 10, 20, and 40 GeV we choose $v/\bar{v} = 3$, 1.8 1.4 and 1.2 respectively. The curves in Fig. 1b are independent of the choice of $m_{\tilde{f}}$ for scalar masses ≥ 20 GeV; both scattering and annihilations are controlled by Z° exchange. It was found[5] that $R > 1$ for $E_\nu < 10$ GeV provided $m_{\tilde{S}} < 40$ GeV. Also $R \simeq 5$ for $m_{\tilde{S}} = 6$ GeV at $E_\nu = 3$ GeV. For both higgsino cases (Figs. 1a,b) secondary neutrinos from $b \to c \to x + \nu$ are important since $\chi\chi \to bb$ is the dominant annihilation channel.

The photino[11,12] is perhaps the most plausible LSP candidate in many models and is the most frequently discussed SUSY dark matter candidate. The relic photino density can be maintained at $\Omega = 1$ by adjusting the squark and slepton masses. For now I will assume that these are all degenerate. As in the symmetric higgsino case, $\tilde{S}$, demanding $\Omega = 1$ results in reduced neutrino fluxes as $m_{\tilde{\gamma}}$ is increased (Fig. 1c). For $m_{\tilde{\gamma}} = 6$, 10, 20 and 40 GeV I have taken $m_{\tilde{f}} \simeq 71$, 88, 117 and 166 GeV to ensure $\Omega = 1$. The flux falls off with increasing m_χ because the elastic scattering cross-section depends on $m_{\tilde{f}}^{-4}$, and because Γ_T also depends on m_χ^{-1}. Quantitatively, the photino results are similar to the symmetric higgsino results.

It is important to stress that in order to test these supersymmetric dark matter candidates, we are really only interested in neutrino events (in underground detectors) with energies in excess of 1 GeV. In addition, because the atmospheric neutrino flux ratio[13] between $\nu_e + \bar{\nu}_e$ and $\nu_\mu + \bar{\nu}_\mu$ is $\simeq .3$ for $E_\nu \simeq 1$-2 GeV and the solar flux ratio is unity, one is experimentally much more sensitive to $\nu_e + \bar{\nu}_e$. To see how close we are to existing data, we can compare these predictions with neutrino observations by the IMB detector when correlated to the direction of the sun[14]. Data with sufficient information (energy and direction) is available for $E_\nu \simeq 400$-2 GeV. When the portion of the data between 1 and 2 GeV is examined a neutrino excess (of $\nu_e + \bar{\nu}_e + \nu_\mu + \bar{\nu}_\mu$) is limited[14] by $R' = \frac{d\Phi}{dE_\nu}\big|_\odot dE_\nu / \frac{d\Phi}{dE_\nu}\big|_{ATM} dE_\nu < 1.3$, (at a 90% confidence level) where $\frac{d\Phi}{dE_\nu}\big|_{ATM}$ is the atmospheric background in the direction of the sun as in the figures but with $\nu_\mu + \bar{\nu}_\mu$ included. R' calculated from the figures must be corrected (to include $\nu_\mu + \bar{\nu}_\mu$) by a factor $2/(1 + \nu_\mu/\nu_e) \simeq .46$ for $E_\nu \simeq 1$-2 GeV. Taking $m_\chi = 6$ GeV im Figs. 1a,b,c one finds $R' = 0.9$, 0.5 and 0.3 respectively, all within a factor of $\simeq 4$ of the present experimental upper limit. Clearly this limit can be improved if more data were available which included higher energies ($E_\nu > 2$ GeV) and/or could distinguish between ν_e and ν_μ. Thus all of the supersymmetric candidates for dark matter are consistent with the present IMB data.

I will now turn to the official topic of my contribution, which is the title of this talk. Although I will discuss very little per se about superstrings[15], I will borrow the necessary essentials in order to extrapolate a lower energy theory from which we can say something about the LSP dark matter candidate. We will begin with E_8 x E_8,[16,17] as the only possible choice for a gauge group consistent with a finite and anomaly free theory which admits chiral fermions. To preserve supersymmetry[16], compactification is assumed to take place on a Calabi-Yau[18] manifold, and the resulting[19] Kahler potential is that of [SU(1,1)/U(1)] X [SU(N,1)/SU(N) X U(1)] supergravity[20]. During compactification, one of the E_8 factors is broken[16,21] to some subgroup of E_6 which is either rank five or six[16,21] or possibly rank four[22], although the latter generally suffers phenomenological problems[23] and I will not discuss it as a possibility any further. The breaking of E_8 is hoped to induce soft supersymmetry breaking through gaugino condensation[24].

To determine which subgroup of E_6 we will wish to concentrate on, it is necessary to first discuss the matter content of these theories. The only non-singlet representation is the 27 (or $\overline{27}$) of E_6[16,21] which has an SO(10) and SU(5) decomposition as 27 = 16 + 10 + 1 = (10 + $\overline{5}$ + 1) + (5 + $\overline{5}$) + 1. The number of fermion generations is the asymmetry between the 27's and $\overline{27}$'s; $N_G = N_{27} - N_{\overline{27}}$ and is given by[16] half of the Euler number of the compactification manifold. The number of $\overline{27}$'s (there is at least one) is also determined topologically by the Hodge number[16] $b_{1,1}$. However, the Hosotani mechanism[25] breaking E_6 during compactification may leave[21] only part of none of the (27 + $\overline{27}$) light. I will assume that no members of these incomplete multiplets are present at low energies.

Symmetry breaking of the rank five or six subgroup of E_6 to the standard model $SU(3)_C$ x $SU(2)_L$ x $U(1)_Y$ takes place conventionally through spontaneous symmetry breaking by the development of a vacuum expectation valve for one or more of the neutral fields contained in the 27. As it turns out, each 27 contains five such fields: $\nu_L(16,5)$, $\nu_L^c(16,1)$, $H(10,5)$, $\bar{H}(10,5)$ and $N(1,1)$. Vevs for H and $\bar{H}$ (the ordinary Higgs doublets of the standard model) are necessary to break $SU(3)_C$ x $SU(2)_L$ x $U(1)_Y \to SU(3)_C$ x $U(1)_{EM}$ leaving only ν_L^c and N. A vev for N is not only possible but essential as it supplies the necessary $\bar{H}\bar{H}$ mixing through the superpotential term $H\bar{H}N$ included in $(27)^3$. If our starting point was rank six, we would now be forced to generate a vev for ν^c as well. In this case, however, phenomenological problems arise[26] because of either the $HL\nu^c$ term in the superpotential (giving rise to large Dirac masses to neutrinos) or the $Dd^c\nu^c$ term (giving rise to D - d mixing) where $D(10,5)$ is an extra down-like quark per generation and $d(16,5)$ is the ordinary down quark. Thus, without the presence of additional fields (e.g., from incomplete $27 + \overline{27}$'s, the rank six groups are not viable candidates and we are left uniquely with rank five of which the minimal gauge group (i.e., with normal color) is $SU(3)_C$ x $SU(2)_L$ x $U(1)$ x $U(1)$. Our starting point, therefore, will be the above low energy gauge group with only three generations of matter fields contained in 27's of E_6[27].

The model contains the following set of new fields: three generations of $D(10,5)$, $D^c(10,5)$, $\nu^c(16,1)$ and $N(1,1)$ and one additional gauge boson Z' coming from the extra $U(1)$ factor as well as their superpartners. What we will be interested in here is examining the cosmological consequences of the model and the potential for testing the model by looking for a new stable particle or new interactions mediated by the extra Z'. The most general renormalizable superpotential which is E_6 invariant is $(27)^3$ and contains the following terms:

$$F = \lambda_1 \bar{H}Le^c + \lambda_2 \bar{H}Qd^c + \lambda_3 HQu^c$$

$$+ \lambda_4 H\bar{H}N + \lambda_5 DD^cN + \lambda_6 DQQ$$

$$+ \lambda_7 D^cu^cd^c + \lambda_8 D^cQL + \lambda_9 Du^ce^c \qquad (7)$$

$$+ \lambda_{10}Dd^c\nu^c + \lambda_{11}HL\nu^c.$$

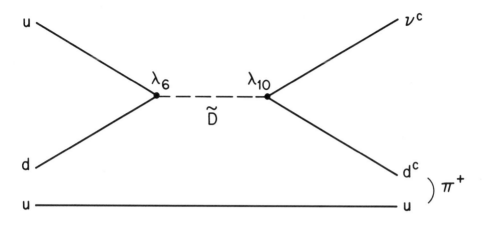

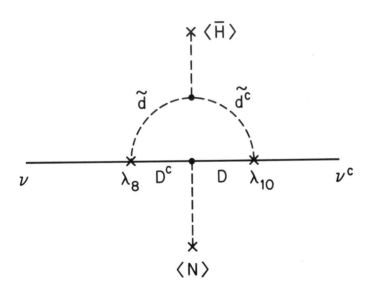

Figure 2: a) Diagram for the proton decay $p \rightarrow \pi^+ \nu^c$; b) one-loop contribution to a Dirac mass for neutrinos.

If all of these terms were present, they would result in serious problems[27,28]. For example, the presence of λ_{11} leads to an unacceptably large Dirac mass for neutrinos and the presence of λ_6 and λ_{10} leads to fast proton decay through $p \rightarrow \pi^+ \nu^c$ (see Fig. 2a) where the scalar D mass is ~ 1 TeV (see below). By the inclusion of discrete symmetries as simple as an R-parity, one can eliminate[27] certain terms such as λ_6, λ_7, λ_8, λ_9 and λ_{11}. In the above superpotential, the first three terms represent the standard model couplings. The λ_4 term was discussed previously and the λ_5 term gives rise to masses for the D-quark $m_D \sim <N> =x \lesssim 1$ TeV due to the vev for N. The terms λ_8, λ_9 and λ_{10} together may lead to acceptable neutrino masses[28,29] (see Fig. 2b).

By defining the lepton number for ν^c and N to be zero and the baryon number for D and D^c to be 1/3. Invariance under R-parity, defined in eq. 3, eliminates precisely the unwanted terms in the superpotential $\lambda_{6-9, \ 11}$ and guarantees the existence of a stable particle as was previously discussed. The LSP will, in general, be a mixture[30] of the neutral R = -1 matter fields $\tilde{H}$, $\tilde{\tilde{H}}$ and $\tilde{N}$ along with the three neutral gauginos $\tilde{\gamma}$, $\tilde{Z}$ and $\tilde{Z}'$. (Although $R(\nu^c) = -1$, it has no mixing with the other R = -1 fields so that one of the above fields must remain stable). As in standard supersymmetric models, we expect that the cosmological limits on the mass density of the Universe to place constraints on the mass of the LSP[11,12,30].

The cosmological mass density of a massive LSP, χ, is determined primarily by the $\chi\chi$ annihilation cross-section ($\rho\propto<\sigma\upsilon>_A^{-1}$). In the low energy superstring model[27] we are considering, the annihilation cross-sections depend primarily on only a single parameter, $m_{1/2}$, the gaugino mass which is the dominant source of supersymmetry breaking[24]. For $m_{1/2}$ in the range 100-1000 GeV, it was found that[30]

$$\Omega_\chi (h_o/.5)^2 = 2^{0 \pm 2} \tag{8}$$

indicating that a substantial contribution to the overall mass density of the Universe. (See Fig. 3). In addition, if we believe that $\Omega_\chi = 1$, then depending on the value of $m_{1/2}$, h_o falls in the range 0.25 - 1 in remarkable agreement with the observed range $1/2 < h_o < 1$. As we will see below, constraints from big bang nucleosynthesis involving the mass of the Z', will further reduce the allowed range of Ω_χ and h_0.

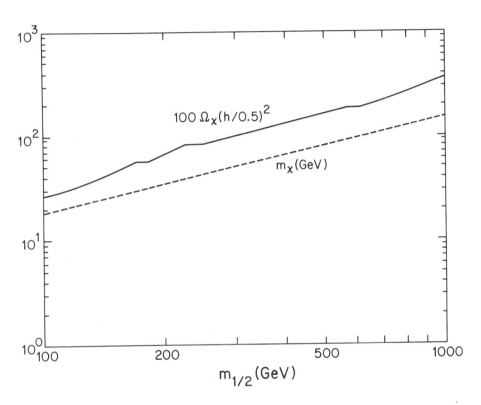

Figure 3: the mass of the LSP, m_χ in GeV and its cosmological relic density $\Omega(h/0.5)^2$ as a function of the gaugino mass $m_{1/2}$.

In the model discussed up to now, either the neutrinos ν_L and ν_L^c are massless[27] or they may have some small Dirac mass[28]. As long as $m_\nu \leq 1$ MeV, both ν and ν^c will contribute to energy density of the Universe and hence the expansion rate at the time of big bang nucleosynthesis. Thus, if all three generations are nearly massless, then the effective number of neutrino states, N_ν, is six, thus possibly in conflict[31] with bounds on N_ν from nucleosynthesis[32,33] $N_\nu < 4$ (if all parameters in the nucleosynthesis calculations such as the abundances of ^{4}He, ^{3}He and the neutron half-life are pushed to their extremes, the limit may be extended[33] to 4.6). But because ν^c interactions are mediated only by Z' rather than Z, the decoupling temperature (the temperature at which interaction rates fall below the expansion rate) of ν^c is greater[34] than ν_L if $M_{Z'} > M_Z$ as it must be. A larger decoupling temperature translates into a lower present day effective temperature[34] for ν^c than ν_L (just as $T_{\nu_L} < T_\gamma$, the temperature of the microwave background radiation. Thus, the effective number of neutrino states becomes

$$N_\nu = 3 + 3(T_{\nu^c}/T_{\nu_L})^3 < 4(4.6) \tag{9}$$

The above constraint implies a lower bound to the mass of the extra neutral gauge boson[33].

$$M_{Z'} > 500 \ (400) \ \text{GeV} \tag{10}$$

In addition, the same argument places a limit on the ratio of vacuum expectation valves, $<0|N|0> \equiv x$ and $<0|H|0> \equiv v$, $x/v > 7(6)$. In this model[27], the supersymmetry breaking scale $m_{1/2}$ is related to x/v, $x/v = (m_{1/2} + 75\text{GeV})/88\text{GeV}$ so that $N_\nu < 4(4.6)$ implies

$$m_{1/2} > 540 \ (450) \ \text{GeV} \tag{11}$$

From Fig. 3, we see that the lower bound on $m_{1/2}$ corresponds to a lower bound on Ωh_o^2

$$\Omega(h_o/.5)^2 \geq 1.6 \ (1.4) \tag{12}$$

and a lower bound on the mass of the LSP

$$m_\chi > 80 \ (70) \ \text{GeV} \tag{13}$$

which for $\Omega_\chi = 1$ corresponds to

$$h_o \geq 0.65 \ (0.6) \tag{14}$$

which is remarkable when compared to the observed range $0.5 \lesssim h_o \leq 1$.

For completeness, we can ask what neutrino signal, if any, one expects from these LSP annihilations in the sun. In Fig. 1d, the differential neutrino flux is again compared to the atmospheric neutrino background. As one can see, there is no useful signature or constraint in this case.

In summary, we have seen that the low energy extrapolation of the superstring predict several new fields contained in 27's of E_6. As in standard supersymmetric models, unbroken R-parity guarantees the existence of one stable particle, the LSP. In the model considered [27], i.e., the low energy subgroup being SU(3) x SU(2) x U(1) x U(1) with only three complete 27's, the LSP is a mixture of the three guaginos $\tilde{W}^3$, $\tilde{B}$ and $\tilde{B}'$ and the Higgsinos $\tilde{H}$, $\tilde{\tilde{H}}$ and $\tilde{N}$. Cosmological constraints from big bang nucleosynthesis imply that $m_\chi \geq 70$ GeV and that if $\Omega_\chi \approx 1$, $h_o \gtrsim 0.6$. Finally, whereas neutrino fluxes produced by $\chi\chi$ annihilations in the sun are large compared with the atmospheric background for light (6GeV $\leq m_\chi \leq 20$ GeV) LSP's, we do not expect a significant signal from our superstring candidate for E_ν 10 GeV.

This work was supported in part by DOE grant DE-AC02-83ER-40105.

References

1) Srednicki, M., these proceedings.

2) Silk, J., Olive, K. A. and Srednicki, M., Phys. Rev. Lett. 55, 257 (1985).

3) Srednicki, M., Olive, K. A., and Silk, J., Nucl. Phys. B (in press), (1986).

4) Gaisser, T. K., Steigman, G., and Tilav, S., Bartol preprint 86-42 (1986).

5) Hagelin, J., Ng, K.-W. and Olive, K. A., UMN preprint UMN-TH-566/86 (1986).

6) Press, W. H. and Spergel, D. N., Ap. J. 296, 679 (1985).

7) Steigman, G., Sarazin, L. L., Quintana, H. and Faulkner, J., Ap. J. 83, 1050 (1978); Spergel, D. N., and Press, W. H., Ap. J. 294, 663 (1985).

8) Krauss, L., Freese, K., Spergel, D. N., and Press, W. H., Ap. J. 299 1001 (1986).

9) Ibanez, L. E., Phys. Lett. 137B, 160 (1984); Hagelin, J., Kane, G., Nucl. Phys. B241, 638 (1984).

10) Gelmini, G., these proceedings; Ahlen, S. P., Avignone, F. T., Brodzinski, R. L., Drukier, A. K., Gelmini, G. and Spergel, D. N., CFA preprint 2292 (1986).

11) Goldberg, H., Phys. Rev. Lett. 50, 1419 (1983); Krauss, L. M., Nucl. Phys. B227, 556 (1983).

12) Ellis, J., Hagelin, J., Nanopoulos, D. V., Olive, K. A. and Srednicki, M., Nucl. Phys. B238, 453 (1984).

13) Gaisser, T. K., Stanev, T., Bludman, S. A. and Lee, H., Phys. Rev. Lett. 51, 223 (1983); Perkins, D. H., Ann. Rev. Nucl. Part. Sci. 34, 1 (1984).

14) LoSecco, J., private communication 1986.

15) Schwarz, J. H., Phys. Rep. 89 (1982) 223; Green, M. B., Surv. High Energy Phys. 3 (1983) 127.

16) Candelas, P., Horowitz, G. T., Strominger, A. and Witten, E., Nucl. Phys. B258 (1985) 46.

17) Green, M. B. and Schwarz, J. H., Phys. Lett. 149B (1984) 117 and 151B (1985) 21.

18) Calabi, E., in Algebraic Geometry and Topology: A Symposium in Honor of S. Lefschetz (Princeton University Press, 1957) p. 78; Yau, S. T., Proc. Nat. Acad. Sci. 74 (1974) 1798.

19) Witten, E., Phys. Lett. 155B (1985) 151; Ellis, J., Gomez, C. and Nanopoulos, D. V., Phys. Lett. 171B (1986) 203.

20) Cremmer, E., Ferrara, S., Kounnas, C. and Nanopoulos, D. V., Phys. Lett. 133B (1983) 61; Ellis, J., Lahanas, A. B., Nanopoulos, D. V., and Tamvakis, K., Phys. Lett. 134B (1984) 429, Ellis, J., Kounnas, C., Nanopoulos, D. V., Nucl. Phys. B241 (1984) 406 and B247 (1984) 373 and Phys. Lett. 143B (1984) 410; Ellis, J., Enqvist, K. and Nanopoulos, D. V., Phys. Lett. 147B (1984) 94 and 151B (1985) 357.

21) Witten, E., Nucl. Phys. B258 (1985) 75; Dine, M., Kaplanousky, V., Mangano, M., Nappi, C., and Seiberg, N., Nucl. Phys. B259 (1985) 519.

22) Dixon, L., Harvey, J. A., Vafa, C., and Witten, E., Nucl. Phys. B261 (1985) 678 and Princeton University preprint (1986).

23) Ellis, J., Enqvist, K., Nanopoulos, D. V., Olive, K. A., Quiros, M. and Zwirner, F., Phys. Lett. 176B, 403 (1986).

24) Derendinger, J.-P., Ibanez, L. and Nilles, H.-P., Phys. Lett. 155B (1985) 65; Dine, M., Rohm, R., Seiberg, N. and Witten, E., Phys. Lett. 156B (1985) 55; Cohen, E., Ellis, J., Gomez, C. and Nanopoulos, D. V., Phys. Lett. 160B (1985) 62.

25) Hosotani, Y., Phys. Lett. 126B (1983) 309.

26) Campbell, B., Ellis, J., Gaillard, M. K., Nanopoulos, D. V. and Olive, K. A., CERN preprint Th-4469/86 (1986).

27) Ellis, J., Enqvist, K., Nanopoulos, D. V and Zwirner, F., CERN preprint Th-4323/85 (1985).

28) Campbell, B., Ellis, J., Enqvist, K., Gaillard, M. K. and Nanopoulos, D. V., CERN preprint TH-4473 (1986).

29) Masiero, A., Nanopoulos, D. V. and Sanda, A. I., Rockefeller preprint RU86/B/158.

30) Campbell, B., Ellis, J., Enqvist, K., Hagelin, J., Nanopoulos, D. V. and Olive, K. A., Phys. Lett. 173B (1986) 270.

31) Ellis, J., Enqvist, K., Nanopoulos, D. V. and Sarkar, S., Phys. Lett. 167B, 457 (1986).

32) Steigman, G., Schramm, D. N. and Gunn, J. E., Phys. Lett. 66B 202 (1977); Olive, K. A., Schramm, D. N., Steigman, G., Turner, M. S. and Yang, J., Ap. J. 246, 557 (1981); Yang, J., Turner, M. S., Steigman, G., Schramm, D. N. and Olive, K. A., Ap. J. 281 493 (1984).

33) Steigman, G., Olive, K. A., Schramm, D. N. and Turner, M. S., Phys. Lett. 176B, 33 (1986).

34) Steigman, G., Olive, K. A. and Schramm, D. N., Phys. Rev. Lett. 43 (1979) 239; Olive, K. A., Schramm, D. N. and Steigman, G., Nucl. Phys. B180 (1981) 497.

On the Present Mass Density

of Relic Photinos

Keith A. Olive

School of Physics and Astronomy

University of Minnesota

Minneapolis, MN 55455

Mark Srednicki

Department of Physics

University of California

Santa Barbara, CA 93106

Joe Silk

Department of Astronomy

University of California

Berkeley, CA 94720

Abstract

We calculate the minimum mass density of relic photinos assuming two choices for slepton masses: 1) $m_{\tilde{\mu}} = m_{\tilde{\tau}} = 23$ GeV and $m_{\tilde{e}}(m_{\tilde{\gamma}})$ given by the latest experimental limits and 2) $m_{\tilde{\mu}} = m_{\tilde{\tau}} = m_{\tilde{e}}(m_{\tilde{\gamma}})$. We have taken all squark masses to be 60 GeV. Any other choice of scalar masses consistent with experimental limits results in a larger present day density of photinos.

Although it is likely that some form of non-baryonic dark matter must be present with a fairly large abundance, its identity remains completely unknown. There exists, of course, a lengthy list of dark matter candidates which include massive neutrinos, axions, supersymmetric particles such as photinos, higgsinos or sneutrinos, etc. The requirement that these particles are stable (that their lifetimes be longer than the age of the Universe) casts some doubt on the naturalness of massive neutrinos ($m \gtrsim \mathcal{O}(\text{GeV})$) as cold dark matter candidates. An attractive feature of the supersymmetric candidates is that one of them is predicted to be stable due to an unbroken R-parity. In almost all supersymmetric models, R-parity defined by

$$R = (-1)^{2S + L + 3B}$$

remains unbroken. "Normal particles such as the quarks, leptons, gauge and higgs bosons all have R = +1, while "supersymmetric" particles all have R = -1. Thus the lightest R = -1 state (LSP) must be stable, making it an excellent candidate for the dark matter. For the rest of this discussion, we will consider the photino as the LSP.

Despite its stability, any dark matter candidate must be present with a sufficient abundance in order to solve a growing list of "dark matter problems".[1] The present abundance of photinos[2],[3] depends primarily on the annihilation cross-section at the time of freezeout (when the rate of annihilations falls below the expansion rate) $\Gamma_f \sim m_{\tilde{\gamma}}/20$

$$\lambda_{\tilde{\gamma}} = \rho_{\tilde{\gamma}}/\rho_c = (1.07 \times 10^{-10})(T_{\tilde{\gamma}}/T_{\gamma})^3 (\frac{T_{\gamma}}{2.8})^3 N_f^{1/2} (\frac{0.5}{h_0})^2 \frac{\text{GeV}^{-2}}{(ax_f + \frac{b}{2}x_f^2)} \quad (2)$$

where $\rho_c = 1.88 \times 10^{-29} h_0^2$ g cm^{-3}, N_f is the number of particle degrees of freedom at freezeout, $x_f \equiv T_f/m_{\tilde{\gamma}}$, and the factor $(T_{\tilde{\gamma}}/T_{\gamma})^3$ takes into account any dilution[4] of the number of photinos between T_f and today. The annihilation cross-section is defined as

$$\langle \sigma v \rangle_A = a + bx \tag{3}$$

where

$$a = 8\pi\alpha^2 \Sigma \; Q_f^4 \; \beta_f \; m_f^2/m_{sf}^4$$

$$b = 16\pi\alpha^2 \Sigma \; Q_f^4 \beta_f \delta_f m_\gamma^2/m_{ff}^4$$

$$\beta_f = (1 - m_f^2/m_\gamma^2)^{1/2}$$

$$\delta_f = 1 - (17 - 3\beta_f^{-2})m_f^2/8m_\gamma^2 \tag{4}$$

where α is the fine structure constant, Q_f is the electric charge of the fermion f and m_{sf} is the mass of the scalar partner of f. Clearly by adjusting the unknown values of m_{sf} and hence a and b one can get any desired value of $\Omega_{\tilde\gamma}$ in Eq. (2).

It was shown[5] however, that by using theoretical relations among the masses of the photino, gluino, squarks and sleptons, that there was in fact a lower bound to $\Omega_{\tilde\gamma}$, $\Omega_{\tilde\gamma} \gtrsim 0.01 \; (0.5/h_0)^2$. Thus making photinos a non-negligible component in the Universe. Here, we will take a somewhat different point of view. By considering only the present experimental lower limits to the sfermion masses m_{sf}, we can derive a lower limit to $\Omega_{\tilde\gamma}$ as a function of $m_{\tilde\gamma}$.

We will consider two cases:

a) We take as the lower bounds to the squark mass[6], $m_{\tilde q} > 60$ GeV, while for the sleptons, we take $m_{\tilde\mu} = m_{\tilde\tau} > 23$ GeV and $m_{\tilde e} > 23$ GeV when $m_{\tilde\gamma} > 13$ GeV and $m_{\tilde e} > 84$ GeV when $m_{\tilde\gamma} < 3$ GeV, by combining the ASP, MAC and CELLO data[7].

b) We take a perhaps more reasonable bound from the theoretical point of view that $m_{\tilde\mu} = m_{\tilde\tau} \geq m_{\tilde e}$ and using the above experimental limit to $m_{\tilde e}$. In the Figure we have plotted the quantity $\Omega_{\tilde\gamma}(h_0/0.5)^2$ as a function of $m_{\tilde\gamma}$ for the two cases.

From the Figure, we see that for $\Omega (h_0/0.5)^2 < 1$, the absolute experimental bounds only require $m_{\tilde\gamma} \geq 2$ GeV, while the theoretically reasonable bounds require $m_{\tilde\gamma} > 5.8$ GeV, this is comparable to the preferred value for the photino mass of ~ 3 GeV for producing large fluxes of antiprotons from photino annihilations in the halo[8]. In addition, we see that for $m_{\tilde{\gamma}} < 10$ GeV, the photino is necessarily an important cosmological constituent with $\Omega_{\tilde{\gamma}}(h_0/0.5)^2 > 0.1$. We have ended the curve at $m_{\tilde{\gamma}} = 23$ GeV because above the 23 GeV one has the possibility that $m_{\tilde{e}} \gtrsim m_{\tilde{\gamma}}$ and the experimental bounds do not help in determining a lower bound to $\Omega_{\tilde{\gamma}}$.

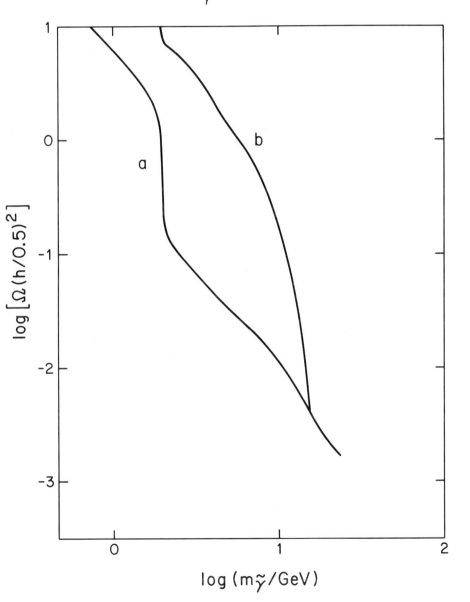

References

1) See eq. Schramm, D. N., Nucl. Phys. B252, 53 (1985).

2) Goldberg, H., Phys. Rev. Lett. 50, 1419 (1983);
   Krauss, L. M., Nucl. Phys. B227, 556 (1983).

3) Ellis, J., Hagelin, J. S. Nanopoulos, D. V., Olive, K. A. and
   Srednicki, M., Nucl. Phys. B238, 453 (1984).

4) Steigman, G., Olive, K. A., and Schramm, D. N., Phys. Rev. Lett.
   43, 239 (1979); Olive, K. A., Schramm, D. N. and Steigman, G.,
   Nucl. Phys. B180, 497(1981).

5) Ellis, J., Hagelin, J. S. and Nanopoulos, D. V., Phys. Lett.
   159B, 26 (1985).

6) Barnett, R. M., Haber, H. E. and Kane, G. L., Nucl. Phys. B267,
   625 (1986).

7) Davier, M., in the Proceedings of the 23rd Int. Conf. on High
   Energy Physics, Berkeley, CA, July 1986.

8) Silk, J. and Srednicki, M. Phys. Rev. Lett. 53, 624 (1984).

The Effects of Resonant Neutrino Oscillations on the Solar Neutrino Experiments[1]

Stephen J. Parke

Fermi National Accelerator Laboratory

P.O. Box 500, Batavia, Illinois, 60510

Abstract

Analytic results are derived for the electron neutrino survival probability after passage through a resonant oscillation region. This survival probability together with a sophisticated model of the production distribution of the solar neutrino sources and the solar electron number density are used to study the effects of resonant neutrino oscillation in the solar interior on the current and proposed solar electron neutrino experiments.

Recently, Mikheyev and Smirnov[1] have shown that the matter neutrino oscillations of Wolfenstein[2] can undergo resonant amplification in the solar interior thereby reducing the flux of electron neutrinos emerging from the Sun. This mechanism may be the solution to the solar neutrino puzzle[3,4]. Subsequently, Bethe[5] and others[6] have refined and restated the Mikheyev and Smirnov idea, pointing out that there are three general regions of parameter space in which the solar electron neutrino flux is sufficiently reduced. In this paper, I report an analytic result[7] for the electron neutrino survival probability after passage through a resonant oscillation region. Then, I outline a calculation[8] which uses this result, together with a relatively sophisticated solar model for the production distribution of solar neutrino sources and the solar electron number density, to generate contour plots of electron neutrino capture rates in the mass difference squared - vacuum mixing angle plane, for both chlorine (^{37}Cl) experiment and the proposed gallium (^{71}Ga) detector.

If neutrinos are massive then the flavor and mass eigenstates are not necessarily identical, however a general neutrino state can always be written in the flavor basis[9],

$$|\nu(t)\rangle = c_e(t)\,|\nu_e\rangle + c_x(t)\,|\nu_x\rangle. \tag{1}$$

In the ultra-relativistic limit, the evolution of this general neutrino state, in matter, is described by the following Schrodinger-like equation[2],

$$i\frac{d}{dt}\begin{pmatrix} c_e \\ c_x \end{pmatrix} = \frac{\Delta_N}{2}\begin{pmatrix} -\cos 2\theta_N & \sin 2\theta_N \\ \sin 2\theta_N & \cos 2\theta_N \end{pmatrix}\begin{pmatrix} c_e \\ c_x \end{pmatrix}. \tag{2}$$

With Δ_N and θ_N determined by

$$\Delta_N \cos 2\theta_N = \frac{\delta m^2}{2k}\cos 2\theta_0 - \sqrt{2}G_F N_e,$$

$$\Delta_N \sin 2\theta_N = \frac{\delta m^2}{2k}\sin 2\theta_0$$

where $\delta m^2 \equiv (m_2^2 - m_1^2)$, m_i are the neutrino masses, k is the neutrino energy, θ_0 is the vacuum mixing angle, G_F is the Fermi constant and N_e is the electron number density. The constraints $\delta m^2 > 0$ and $\theta_0 < \pi/4$ are assumed.

At an electron density, N_e, the matter mass eigenstates are

$$|\nu_1, N\rangle = \cos\theta_N\,|\nu_e\rangle - \sin\theta_N\,|\nu_x\rangle$$

$$|\nu_2, N\rangle = \sin\theta_N\,|\nu_e\rangle + \cos\theta_N\,|\nu_x\rangle \tag{3}$$

which have eigenvalues $E_1 = -\Delta_N/2$ and $E_2 = \Delta_N/2$. At resonance, the electron density is given by $N_e^{res} = \delta m^2 \cos 2\theta_0/2\sqrt{2}kG_F$, and the matter mixing angle $\theta_N^{res} = \pi/4$. Above resonance, θ_N satisfies $\pi/4 < \theta_N \leq \pi/2$.

For a constant electron density these matter mass eigenstates evolve in time by the multiplication of a phase factor. Also, for a slowly varying electron density, these states evolve independently in time; i.e. $e^{-i\int^t E_1 dt}\,|\nu_1, N(t)\rangle$ and $e^{-i\int^t E_2 dt}\,|\nu_2, N(t)\rangle$ are the adiabatic states. Therefore, it is convenient to use these states, as the basis states, in the region for which there are no transitions (away from the resonance region). As a neutrino goes through resonance these adiabatic states maybe mixed, but on the other side of resonance, the neutrino state can still be written as a linear combination of these states. That is, a basis state produced at time t, going through resonance at time t_r, and detected at time t' is described by

$$e^{-i\int_t^{t_r} E_1 dt}\,|\nu_1, N(t)\rangle \rightarrow$$

$$a_1\,e^{-i\int_{t_r}^{t'} E_1 dt}\,|\nu_1, N(t')\rangle + a_2\,e^{-i\int_{t_r}^{t'} E_2 dt}\,|\nu_2, N(t')\rangle$$

or

$$e^{-i \int_t^{t_r} E_2 dt} \; |\nu_2, N(t)\rangle \; \rightarrow$$

$$-a_2^* \, e^{-i \int_{t_r}^{t'} E_1 dt} \; |\nu_1, N(t')\rangle \; + \; a_1^* \, e^{-i \int_{t_r}^{t'} E_2 dt} \; |\nu_2, N(t')\rangle$$

where a_1 and a_2 are complex numbers such that $|a_1|^2 + |a_2|^2 = 1$. The relationship between the coefficients, for these two basis states, is due to the special nature of the wave equation, eqn(2). The phase factors have been chosen so that coefficients a_1 and a_2 are characteristics of the transitions at resonance and are not related to the production and detection of the neutrino state.

The detection averaged electron neutrino survival probability is easily calculated as

$$P_{\nu_e}(t) \;=\; \frac{1}{2} + \frac{1}{2}(|a_1|^2 - |a_2|^2)\cos 2\theta_N \cos 2\theta_0$$

$$- \; |a_1 a_2| \sin 2\theta_N \cos 2\theta_0 \cos(\int_{t_r}^t \Delta_N dt \; + \; \omega))$$

with $\omega = \arg(a_1 a_2)$. The last term shows that the phase of the neutrino oscillation at the point the neutrino enters resonance can substantially effect this probability. Therefore, we must also average over the production position, to obtain the fully averaged electron neutrino survival probability[7,10] as

$$\overline{P_{\nu_e}} \;=\; \frac{1}{2} \; + \; (\frac{1}{2} - P_x)\cos 2\theta_N \cos 2\theta_0 \tag{4}$$

where $P_x = |a_2|^2$, the probability of transition from $|\nu_2, N >$ to $|\nu_1, N >$ (or vice versa) during resonance crossing. The non-resonant crossing case is trivially obtained by setting $P_x = 0$.

To calculate the probability, P_x, the approximation that the density of electrons varies linearly in the transition region is used. That is, a Taylor series expansion is made about the resonance position and the second and higher derivative terms are discarded. In this approximation the probability of transition between adiabatic states was calculated by Landau and Zener. This is achieved by solving the Schrodinger equation, eqn(2), exactly in this limit. Applying their result to the current situation[7,11] gives

$$P_x = \exp\left[-\frac{\pi \sin^2 2\theta_0}{2 \cos 2\theta_0} \frac{\delta m^2/2k}{|\vec{n} \cdot \nabla \ln N_e|_{res}} \right] \tag{5}$$

where the unit vector, $\vec{n}$, is in the direction of propagation of the neutrino. Eqn(4) and (5) demonstrate that only the electron number density, at production, and the logarithmic derivative of this density, at resonance, determine the electron neutrino survival probability. For a discussion of the range of validity of this approximation see ref(7).

Before applying these results to the solar model in detail, let us first consider an exponential electron number density profile which is a good approximation for the solar interior except near the center. In figure (1), I have plotted the electron neutrino survival probability contours at the earth in the $\delta m^2 / 2\sqrt{2} k G_F N_c$ versus $\sin^2 2\theta_0 / \cos 2\theta_0$ plane for such an exponential density profile. Here, the Solar central electron number density, N_c, is also the number density at the point where the neutrinos are produced. This plot depends only on the properties of the sun and this dependency is through the combination $R_s N_c$, where R_s is the scale height. For this figure, I have used an N_c corresponding to a density of 140g cm^{-3} and $Y_e = 0.7$ and a scale height R_s of 0.092 times the radius of the sun.

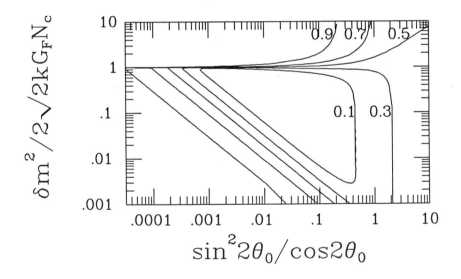

Fig. (1): Electron neutrino survival probability contours for an exponential solar electron density profile and an electron neutrino produced at the center of the Sun.

Above the line $\delta m^2/2\sqrt{2}kG_F N_c = 1/\cos 2\theta_0$ in this plot, the neutrinos never cross the resonance density on there way out of the sun. Here, the probability of detecting an electron neutrino is close to the standard neutrino oscillation result. Below this line, the effects of passing through resonance comes into play. Inside the 0.1 contour "triangle", there is only a small probability of transitions between the adiabatic states as the neutrino passes through resonance. To the right of this contour triangle, the probability of detecting a neutrino grows, not because of transitions, but because both adiabatic states have a substantial mixture of electron neutrino at zero density. To the left and below the 0.1 contour triangle, the probability grows because here there are significant transitions between the adiabatic states as the neutrino crosses resonance.

More precisely, the solar electron neutrino capture rate for a detector charac-terized by a electron neutrino capture cross section, $\sigma(E)$, and energy threshold E_0, is

$$\sum_{processes} \int_{E_0}^{\infty} \frac{d\Phi_\nu}{dE} \sigma(E) dE. \tag{6}$$

The sum is taken over all neutrino sources in the Sun and $d\Phi_\nu/dE$ is the differ-ential electron neutrino flux of a given source at the earth's surface. To include the reduction in the electron neutrino flux from the Sun due to resonant neutrino oscillations, the differential electron neutrino flux for each process was calculated as

$$\frac{d\Phi_\nu}{dE} \propto W(E) \int_{sun} dV\, \overline{P_{\nu_e}} \frac{df}{dV} \tag{7}$$

where $W(E)$ is the standard weak interaction energy distribution for the neutri-nos of a given process and df/dV is the fraction of the standard solar model flux coming from a given solar volume element for this process. The solar electron number density profile, $\rho Y_e/m_N$, and the values of df/dV for the various pro-cesses were taken from Bahcall's solar model[12], see figure (2). We have assumed that the spatial distributions for pep and CNO neutrinos are given by those for pp and [8]B neutrinos, respectively[13]. We normalize $d\Phi_\nu/dE$ for each process by demanding that the energy and solar volume integrations of eqn.(6) yield the capture rates quoted by Bahcall when $\overline{P_{\nu_e}} = 1$.

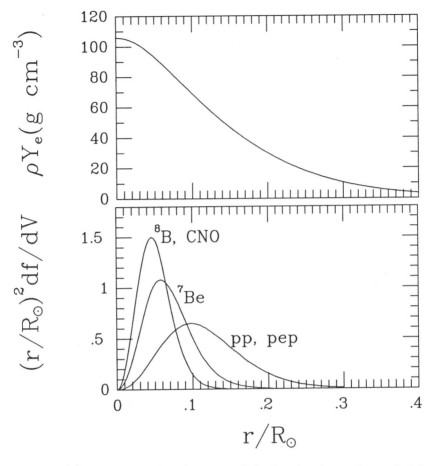

Fig. (2):Electron number density and the fractional neutrino emissivity per shell verus the radial distance from the solar center.

The cross sections, $\sigma(E)$, used for the ^{37}Cl and ^{71}Ga detectors, whose thresholds are 814 and 236 keV respectively, are given in figure (3). The ^{37}Cl cross section is derived from the data of Bahcall[12] and the ^{71}Ga cross section is a fit to the low energy calculation of Bahcall[14] and the higher energy calculations of Grotz, Klapdor, and Metzger[15]. In Table I, we list two sets of expected capture rates for both the chlorine and gallium experiments and the maximum neutrino energy for each solar neutrino source. Model A is taken from the values of Bahcall et al.[4] and Model B, reported by Bahcall[16], reflects recent changes in the expected solar neutrino capture rate. The most important change being in the estimation of the Sun's opacity which alters the solar temperature profile. A comparison between these two models demonstrates the insensitivity of the allowed region of parameter space to small changes in the solar model. The value of 16 SNU for the ^{8}B rate in Model A for the gallium experiment is an average of the new predictions of Grotz *et al.*and Mathews *et al.*[17].

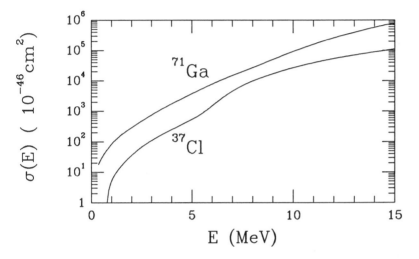

Fig. (3): The cross sections for the Chlorine and Gallium detectors as a function of energy.

Table I: Neutrino Sources and Capture Rates for Two Solar Models

Process	E_ν^{max}(MeV)	Chlorine (SNU)		Gallium (SNU)	
		Model A	Model B	Model A	Model B
^{8}B	14.06	4.3	5.75	16.0	18.0
^{7}Be	0.861(90%)				
	+ 0.383(10%)	1.0	1.1	27	34
p-p	0.420	0	0	70	70
pep	1.44	0.23	0.20	2.5	3.0
^{13}N	1.199	0.08	0.10	2.6	4.0
^{15}O	1.732	0.26	0.35	3.5	6.0
Total		5.9	7.5	122	135

160

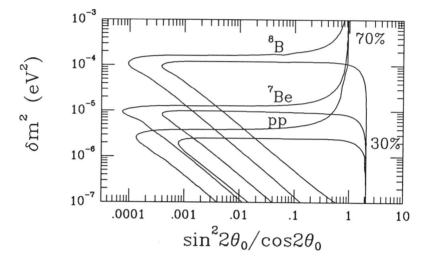

$$\delta m^2 \ (eV^2)$$

$$\sin^2 2\theta_0 / \cos 2\theta_0$$

Fig. (4): The 70% and 30% electron neutrino rate contours for the principal solar neutrino sources in the Gallium experiment.

From Table I we can see that the ^{8}B, ^{7}Be and pp neutrinos account for about 90% of the capture rate for both the ^{37}Cl and ^{71}Ga experiments. Therefore by understanding the effect of resonant neutrino oscillation on these processes we can understand the effect on both of these experiments. In figure (4) I have plotted the 30% and 70% electron neutrino capture rate contours for the Gallium experiment for the three main types of neutrinos emitted from the Sun. For the Chlorine experiment the ^{8}B and ^{7}Be contours are indistinguishable from those in figure (4) because of the similarity of the cross sections energy dependence for these experiments. The total electron neutrino capture rates are obtained by an appropiately weighted superposition of these contours.

In figures 5 and 6, we present electron neutrino capture rate contours (iso-SNU contours) for the ^{37}Cl and ^{71}Ga experiments as a function of δm^2 and $\sin^2 2\theta_0 / \cos 2\theta_0$ for the two solar models discussed earlier. The 1σ deviations from the Davis et al.[3] result of 2.1 SNU are the 2.4 and 1.8 iso-SNU contour lines in fig. 5 The similarity of the shape of these plots for the two solar models reflects the insensitivity of the resonant oscillation process to small changes in the structure of the Sun. However, the position of individual contours does change, due to changes in the contributions from the individual neutrino sources.

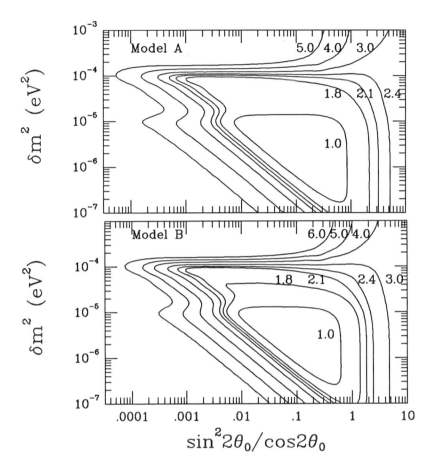

Fig. (5): Iso-SNU contours for the ^{37}Cl experiment for the solar models listed in Table I. The contours are labeled with their the corresponding SNU values.

The generic structure of these total SNU plots is due to the superposition of triangular iso-SNU contours associated with each individual neutrino source contributing to a given total SNU value. These individual contours owe their shape to the appropriate iso-probability contour, figure (1), and their position is determined by the typical energy scale and production electron density of the individual neutrino source, figure (4). For each neutrino source the resonance mechanism becomes important, provided $\theta_0 > 0.01$, as soon as δm^2 becomes small enough so that the average resonant electron density for that source is less than the solar electron density at the production site. This occurs when δm^2 is approximately equal to 1.5×10^{-4}, 1.2×10^{-5}, and 3.7×10^{-6} eV^2 for the ^{8}B, ^{7}Be and pp neutrinos respectively. Below these values the individual neutrino sources have contours which are diagonals of slope minus one coming from the form of the transition probability between adiabatic states, eqn(5). The intersection of these diagonal lines with the turning on of resonance for ^{8}B, ^{7}Be and pp is responsible for the shoulders at small $\sin^2 2\theta_0/\cos 2\theta_0$ in the contour plots. The vertical sections of the contours, at large θ_0, occurs because for large θ_0 both adiabatic states have a large component of electron neutrino.

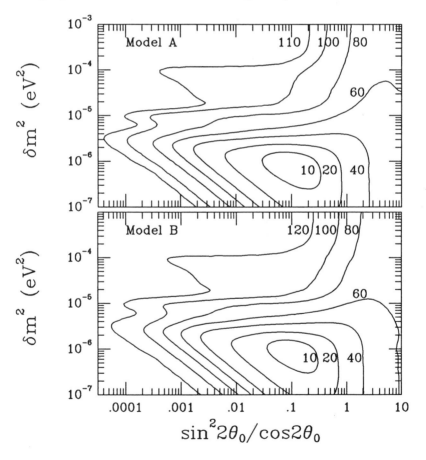

Fig. (6): Iso-SNU contours for a ^{71}Ga detector for the solar models listed in Table I. The contours are labeled with their the corresponding SNU values.

From fig. 6, we see that the results of the ^{71}Ga experiment can range from 10 to 120 SNU and still be compatible with the ^{37}Cl experiment. In general, a given gallium contour crosses the 2.1 ± 0.3 chlorine contour at least twice and therefore the results of the ^{71}Ga experiment will leave a two-fold degeneracy in $(\delta m^2, \theta_0)$-space. If one accepts the theoretical prejudice against large vacuum angles provided by see-saw models[18], this degeneracy is removed. Unfortunately, the degeneracy is continuous for that region of parameter space corresponding to a ^{37}Cl rate of 2.1 ± 0.3 SNU and a ^{71}Ga rate greater than 100 SNU. In this region *only* the ^{8}B neutrinos are effected by the resonance phenomena. Also, in this region of parameter space the two experiments will not be able to distinguish between a small temperature change at the solar core and the resonant neutrino oscillation mechanism. This is due to the relatively strong temperature dependence of the ^{8}B neutrino flux[19]. It is only when the ^{71}Ga SNU rate is depleted below that of merely removing the ^{8}B component (*i.e.*, appreciably less than 110 SNU), so that reduction of the less temperature sensitive neutrinos (^{7}Be and pp) becomes necessary, that the resonant oscillation mechanism becomes a likely solution to the solar neutrino problem.

Many thanks go to my collaborator Terry Walker.

Fermilab is operated by Universities Research Association Inc. under contract with the United States Department of Energy.

References

1. S.P. Mikheyev and A.Yu. Smirnov, *10th International Workshop on Weak Interactions and Neutrinos*, Savonlinna, Finland(1985); *Nuovo Cimento* **C9**, 17(1986).

2. L. Wolfenstein, *Phys. Rev.* **D17**, 2369(1978); *Phys. Rev.* **D20**, 2634(1979).

3. R. Davis, D.S. Harmer, and K.C. Hoffman, *Phys. Rev. Lett.* **20**, 1205(1968).

4. J.N. Bahcall, B.T. Cleveland, R. Davis, and J.K Rowley, *Astrophys. J.* **292**, L79(1985).

5. H.A. Bethe, *Phys. Rev. Lett.* **56**, 1305(1986).

6. A. Messiah and S.P. Rosen and M. Spiro, *1986 Massive Neutrinos in Astrophysics and in Particle Physics*, Tignes, January 1986; S.P. Rosen and J.M. Gelb, *Phys. Rev.* , **D34**, 969(1986); E.W. Kolb, M.S. Turner, and T.P. Walker, *Phys. Lett.*, **B175**, 478(1986); V. Barger, R.J.N. Phillips, and K. Whisnant, *Phys. Rev.* , D34, 980(1986); J. Bouchez, M. Cribier, J. Rich, M. Spiro, D. Vignard and W. Hampel, DPhPE 86-10, May 1986.

7. S.J. Parke, *Phys. Rev. Lett.* **57**, 1275(1986).

8. S.J. Parke and T.P. Walker, *Phys. Rev. Lett.* (to be published).

9. The other flavor eigenstate could ν_μ or ν_τ.

10. A. Messsiah, see ref. 6.

11. W.C. Haxton, *Phys. Rev. Lett.* **57**, 1271(1986).

12. J.N. Bahcall, W.F. Huebner, S.H. Lubow, P.D. Parker, and R.K. Ulrich, *Rev. Mod. Phys.* **54**, 767(1982).

13. The CNO distributions are undoubtedly tighter than ^{8}B due to their more sensitive dependence on temperature, but the inaccuracy introduced by this approximation is negligible.

14. J.N. Bahcall, in *Proceedings of the Neutrino Mass Miniconference*, eds. V. Barger and D. Cline, Telemark, Wisconsin(1980).

15. K. Grotz, H.V. Klapdor, and J. Metzinger, *Astron. Astrophys.* **154**, L1(1986).

16. J. N. Bahcall, *International Symposium on Weak and Electromagnetic Interactions in Nuclei*, Heidelberg, July 1986.

17. G.J. Mathews, S.D. Bloom, G.M. Fuller, and J.N. Bahcall, *Phys. Rev.* , **C32**, 796(1985).

18. T. Yanagida, *Prog. Theor. Phys.* **B135**, 66(1978); M. Gell-Mann, P. Ramond, and R. Slansky, in *Supergravity*, eds. P. van Nieuwenhuizen and D. Freedman, (North Holland)(1979).

19. In the case where a ^{71}Ga rate of $\gtrsim$ 100 SNU is measured, a measurement of the ^{8}B solar neutrino spectrum (see Rosen and Gelb ref. 6) or the flavor independent solar neutrino spectrum (see S. Weinberg, *XXIII International Symposium on High Energy Physics*, Berkeley, July 1986) would allow us to distinguish between changes in the solar model and resonant neutrino oscillations.

A SLOW ROLLOVER PHASE TRANSITION
IN THE SCHRÖDINGER PICTURE

So-Young Pi

Department of Physics
Boston University
590 Commonwealth Avenue
Boston, Massachusetts 02215 U.S.A.

ABSTRACT

The present status of our understanding of the slow-rollover transition in the new inflationary universe is reviewed and a time-dependent variational approximation in the Schrödinger picture is proposed as the perturbative scheme for studying the quantum theory of the transition. Validity of the approximation is discussed using a double-well potential, in one-dimensional quantum mechanics.

I. INTRODUCTION

Inflation[1,2] is a simple and powerful cosmological scenario which can resolve several fundamental difficulties in the standard cosmology. However, in order to enhance this idea to a correct cosmological theory, one must establish that observations are consistent with the Zeldovich spectrum of adiabatic density perturbations[3] and that the density parameter Ω is equal to unity; both are definite predictions of inflation. Moreover, one must show that inflation actually occurs by studying detailed quantum dynamics.[4] In this talk, I shall describe some of my collaborative research[5,6] on the quantum evolution of the inflation-driving scalar field in the new inflationary scenario.

An exact calculation of the field theoretic quantum dynamics involves too many degrees of freedom to be reduced to a simple numerical computation. Therefore, in order to study the quantum theory for a scalar field, one must first decide on an approximation scheme, and also find the domain of its validity. My main purpose here is to propose the "time-dependent variational approximation"[7] as the appropriate method for our problem, and to exhibit the domain of its validity by applying it to a one-dimensional quantum mechanical problem which has many features appearing in the inflationary scenario.

To begin, let me describe briefly the standard picture of the new inflationary scenario where inflation occurs during a slow rollover transition.[2] The original scenario is based on the following three assumptions:

i) The initial state of the scalar field ϕ is in thermal equilibrium; the potential for ϕ is extremely flat near $\phi = 0$ at zero temperature and has minima at $\phi = \pm\phi_c$. At extremely high temperatures, $T > T_c$ the effective potential has a unique minimum at $\phi = 0$. Assumption i) implies that $\langle\phi\rangle = 0$ at $T > T_c$, and that the universe is in the high temperature minimum.

ii) As time evolves, the scalar field ϕ stays near $\phi = 0$, until $T << T_c$, *i.e.* supercooling occurs.

iii) Eventually, the slow rollover transition takes place such that the amplitude of ϕ increases slowly with time according to the classical equation of motion.

In a preliminary analysis[5] of the quantum theory of the slow rollover transition, A. Guth and I have shown the validity of assumptions ii) and iii) when one takes the initial state to be thermal, according to assumption i). We constructed an exactly soluble toy model for the behavior of the scalar field in the new inflationary scenario, where the dynamics of the scalar field is described by the following: A de Sitter background metric is taken for all times. (This is a reasonable approximation for most stages of the rollover; however, the question how the metric changes from the radiation dominated phase to the de Sitter phase is not studied.) Also, we took a free field potential, given by $V(\phi) = -\frac{1}{2}\mu^2\phi^2 + cT^2\phi^2$, where $\mu^2 > 0$ and c is a constant of order of the quartic self-coupling constants and couplings with other fields. T represents a background temperature which changes as $1/R$ with time, where R is the cosmic scale factor. $V(\phi)$ changes with time from a stable potential to an unstable one. Our toy model is certainly only an approximation to the realistic one, but we believe that it describes qualitatively the correct physics for a weakly interacting scalar field, except that it cannot describe how the field settles down at the true minimum at $\phi = \phi_c$.

Among other results of our toy model, the following two indicate that when the initial state of ϕ is in thermal equilibrium, assumptions ii) and iii) are valid.

First, when $T \le T_c$, $\langle\phi\rangle = 0$, but $\langle\phi^2\rangle \sim T^2$. However, if $c <<< 1$, this large initial fluctuation decrease with time until $T << T_c$; it does not remain at the large value $\langle\phi^2\rangle \sim T_c^2$. This implies that the system becomes prepared for the slow rollover transition.

Second, for "large" time, which may be defined as the time when the effective wavelength of each momentum mode becomes much larger than the horizon distance, each mode of ϕ obeys a classical equation of motion, but the time at which ϕ begins to roll is described by a classical probability distribution. I shall discuss later, using one-dimensional quantum mechanics, how this classical behavior appears in a quantum roll.

Although we have proposed that our model contains the correct physics of a slow rollover transition, it turns out that assumption i) in the original scenario, *i.e.* that the initial state is thermal, does not seem to be valid. This is due to the fact that density fluctuations[3] predicted by inflation require that the inflation driving scalar field interacts extremely weakly, with coupling constants of order 10^{-12}. A simple calculation shows that such a weakly coupled field cannot be in thermal equilibrium with other fields. One must assume a random non-equilibrium initial configuration, and study how it evolves with time. A purely classical analysis of this problem has been carried out by Albrecht, Brandenberger and Matzner.[4] They conclude that inflation is possible through a slow rollover transition for wide classes of initial configurations for ϕ.

Now I shall discuss how one can study perturbatively the quantum evolution of an interacting scalar field with non-equilibrium initial configurations. First of all, my previous analysis with Guth suggests that one obtains a clearer understanding of the time evolution in a functional Schrödinger picture. The approximation scheme I shall use is the time-dependent variational method developed by Jackiw and Kerman,[7] introduced in Section II. In section III, by applying it to a double welled potential $V(Q) = \frac{\lambda}{24} \left(Q^2 - a^2\right)^2$ in one-dimensional quantum mechanics, I shall exhibit the domain of validity of the approximation and the late time classical behavior of a particle which starts to roll from $Q = 0$, with $\dot{Q} = 0$. Finally, in section IV, I shall generalize the formalism of section II, to a scalar field in a Robertson-Walker metric.

II. TIME-DEPENDENT VARIATIONAL APPROXIMATION

The time-dependent variational principle, posited by Dirac,[8] is an unconventional and novel approach for studying time-dependent quantum systems. We shall review the subject following the work of Jackiw and Kerman.[7].

Following Dirac, one considers time-dependent states $|\psi, t\rangle$ and requires that the time-integrated diagonal matrix element of $i\hbar\partial_t - H$,

$$\Gamma \equiv \int dt \, \langle \psi, t \, | i\hbar\partial_t - H | \, \psi, t \rangle \tag{2.1}$$

be stationary against variation of $|\psi, t\rangle$. Supplemented by appropriate boundary conditions, this provides a derivation of the time-dependent Schrödinger equation.

The quantity Γ is an effective action for a given system described by $|\psi, t\rangle$ and variation of Γ is the quantum analogue of Hamilton's principle. When a specific *Ansatz* is made for the state $|\psi, t\rangle$, the time-dependent Hartree-Fock approximation emerges and this approach is widely used by quantum chemists[9] and nuclear physicists.[10]

Consider one-dimensional quantum mechanical systems with $H = \frac{1}{2m}P^2 + V(Q)$, and suppose that we are interested in the time evolution of a given initial Gaussian wave function in a time-dependent variational approximation. The effective action involves the diagonal matrix element of $i\hbar\partial_t - H$ in a trial wave function, which we take to be the most general Gaussian,

$$\langle Q|\psi, t\rangle_V = N \exp\left[-\frac{1}{2\hbar}(Q - q)^2 B + \frac{i}{\hbar}p(Q - q) \right] \, . \tag{2.2}$$

Following Jackiw and Kerman, we parametrize the real and imaginary part of B as

$$B(t) = \frac{1}{2}G^{-1}(t) - 2i\Pi(t) \, . \tag{2.3}$$

The normalization factor N is then $(2\pi\hbar G)^{-1/4}$. The real quantities $p(t)$, $q(t)$, $G(t)$ and $\Pi(t)$ are the variational parameters and we demand that their variations vanish at $t = \pm\infty$.

For potentials $V(Q)$ which are quartic in Q, the explicit form of Γ_V is

$$\Gamma_V(q, p, G, \Pi) = \int_{-\infty}^{\infty} dt \left[p\dot{q} - H_{cl}(q, p) \right.$$

$$\left. + \hbar \left(\Pi\dot{G} - \frac{1}{2}GV^{(2)}(q) - \frac{1}{8}G^{-1} - 2\Pi^2 G \right) - \frac{\hbar^2}{8}G^2 V^{(4)}(q) \right] \tag{2.4}$$

where $H_{cl}(q,p) = \frac{1}{2m}p^2 + V(q)$ and $V^{(n)}(q) \equiv \frac{\partial^n V(q)}{\partial q^n}$. We note that $\Pi(t)$ plays the role of the momentum conjugate to $G(t)$. The four variational equations are then

$$\frac{\delta\Gamma_V}{\delta p} = 0 \rightarrow \dot{q} = \frac{1}{m}p \tag{2.5a}$$

$$\frac{\delta\Gamma_V}{\delta\Pi} = 0 \rightarrow \dot{G} = 4G\Pi \tag{2.5b}$$

$$\frac{\delta\Gamma_V}{\delta q} = 0 \rightarrow \dot{p} = -V^{(1)}(1) - \frac{\hbar}{2}GV^{(2)}(q) \tag{2.5c}$$

$$\frac{\delta\Gamma_V}{\delta G} = 0 \rightarrow \dot{\Pi} = \frac{1}{8}G^{-2} - 2\Pi^2 - \frac{1}{2}V^{(2)}(q) - \frac{\hbar}{4}V^{(4)}(q)G \tag{2.5d}$$

We call the above "time-dependent Hartree-Fock" (HF) equations, because using the Gaussian wave function leads to the approximation in which all n-point expectation values are expressed in terms of one- and two-point functions.

III.   QUANTUM ROLL IN A DOUBLE WELLED POTENTIAL

This Section is part of work I have carried out in collaboration with F. Cooper and P. Stancioff.[6]

Suppose at $t = 0$, a particle is described by a Gaussian wave function centered at $Q = 0$. We are interested in the time-evolution of the particle in a potential, $V(Q) = \frac{\lambda}{24}\left(Q^2 - a^2\right)^2$, in the time-dependent variational approximation.

First, let me present the result in an upside-down harmonic oscillator, $V(Q) = -\frac{1}{2}kQ^2$, $k > 0$, obtained by Guth and myself.[5] Here the time-dependent Schrödinger equation is exactly soluble. The wave function is given by

$$\psi(Q,t) = (2\pi G\hbar)^{-1/4} \exp\left[-\frac{1}{2b^2}\tan\left(\alpha - i\omega t\right)Q^2 \right] \tag{3.1}$$

where $b^2 = \hbar/\sqrt{mk}$, $\omega^2 = k/m$ and α is a real integration constant. Note that b is a natural quantum mechanical scale of this problem.

For large times, the above wave function provides the following information:

i) The probability distribution for Q is given by

$$\langle Q^2 \rangle \longrightarrow \frac{1}{4} \frac{b^2}{\sin 2\alpha} e^{2\omega t} \tag{3.2}$$

i.e. $\sqrt{\langle Q^2 \rangle}$ obeys a classical equation of motion.

ii) Application of the momentum operator to ψ, yields

$$-i\hbar \frac{\partial \psi}{\partial Q} = \left(\frac{i}{2} G^{-1}(t) + 2\Pi(t) \right) Q\psi$$

$$= \left[\sqrt{mk}Q + 0 \left(e^{-2\omega t} \right) \right] \psi \tag{3.3}$$

Note that $\sqrt{mk}Q$ is the classical momentum $p_{\text{cl}} = \sqrt{2m \left(E - V(Q) \right)}$ which would be attained by a *classical* particle at Q which rolled from rest at $Q = 0$ at total energy $E = 0$.

iii) The commutator $[Q, P]$ is negligible if

$$\sqrt{mk}Q^2 >> \hbar \quad , \quad i.e. \quad Q^2 >> b^2 \quad . \tag{3.4}$$

Note however that the wave function is definitely not sharply peaked about any particular classical trajectory. Rather, at large times the system is described by a classical probability distribution,

$$f(Q, P, t) = |\psi(Q, t)|^2 \delta(P - p_{\text{cl}}) \quad . \tag{3.5}$$

The function f obeys a classical evolution equation.

Now let us turn to the double well potential. In the variational approximation (or HF approximation), our trial wave function for this problem is

$$\psi_V(Q, t) = (2\pi G)^{1/4} \exp \left[-\frac{1}{2} Q^2 \left(\frac{1}{2} G^{-1}(t) - 2i\Pi(t) \right) \right] \tag{3.6}$$

(We have set $\hbar = 1$.) A natural quantum mechanical length scale b may be defined as

$$b^2 \equiv \frac{1}{\sqrt{mk}} \quad ; \quad k \equiv \left| V^{(2)}(0) \right| = \frac{1}{6} \lambda a^2 \quad . \tag{3.7}$$

We shall choose our initial width of the Gaussian to be

$$G_0 = \frac{1}{2} b^2 = \left(\frac{3}{2\lambda a^2} \right)^{1/2} \tag{3.8}$$

Since our initial conditions are fixed, one can perform a numerical calculation to solve Eqs. (2.5).

Let me now discuss the domain of validity of the time-dependent HF approximation. In this approximation, some properties of $G(t)$ can be determined exactly and in particular its maximum value is found to be

$$G_{\max} = \frac{2}{3}a^2 \ , \tag{3.9}$$

i.e. $\sqrt{G}$ never spreads sufficiently to reach the minima at $R = \pm a$. This premature turning point shows that the approximation fails near the bottom of the two wells. This failure is due to the fact that the approximation is based on a single Gaussian wave function which cannot describe probability piling up at the two minima. However, as we see in Figs. 1a and 1b, when $G \leq G_{\max}$ the approximation is excellent for all λ in comparison with the exact solution obtained by a numerical calculation in the Heisenberg picture.

Next, we shall discuss the large time behavior of the system. Unlike the upside-down harmonic oscillator whose potential is always unstable, so that Q increases indefinitely, here "large time" is rather limited; it is some intermediate time before $\sqrt{\langle Q^2 \rangle}$ arrives near the minimum. In fact, in the HF approximation, "large time" must be when $G(t) \lesssim \frac{2}{3}a^2$.

From Eq. (3.4), we expect that classical behavior may appear for $\langle Q^2 \rangle >> b^2$. This requires that

$$R \equiv \frac{b^2}{a^2} << 1 \longrightarrow \frac{\sqrt{6}}{\sqrt{m\lambda}\,a^3} << 1 \ . \tag{3.10}$$

Eq. (3.10) implies that for given m and a, λ must be large. One may define a dimensionless coupling constant for this problem in terms of the mass and of the quantum mechanical scale b as

$$\lambda' \equiv mb^6\lambda = \frac{6\sqrt{6}}{\sqrt{m\lambda}\,a^3} \ . \tag{3.11}$$

$\lambda' << 1$ is equivalent to $R << 1$. For our numerical calculation we have chosen $m = 1$, $a = 5$ and two values of λ; $\lambda = 3.84$ and $\lambda = 0.0123$.

The following are the results for small dimensionless coupling constant, $\lambda' = 0.06$, which corresponds to $\lambda = 3.84$:

i) In Fig. 1a we compare $\sqrt{\langle Q^2 \rangle}$ in the exact solution (obtained numerically) to its HF approximation $\sqrt{G(t)}$. We find excellent agreement until $\sqrt{G}$ reaches its premature turning point at $\pm a\sqrt{\frac{2}{3}}$. Despite this, a reasonable result for the oscillation time is obtained in the HF approximation.

ii) Classical behavior of $\langle Q^2 \rangle$ in the late time: Since a classical particle with $Q(0) = P(0) = 0$, and therefore $E = \frac{\lambda}{24}a^4$, will stay at $Q = 0$, we did our computer experiment by placing the classical particle at $Q(0) = \sqrt{G_0}$ and compared this particular classical trajectory $Q_{\text{cl}}(t)$ with the exact $\sqrt{\langle Q^2 \rangle}$ calculation in the Heisenberg picture. In Fig. 2, we find that for $1.5 <$

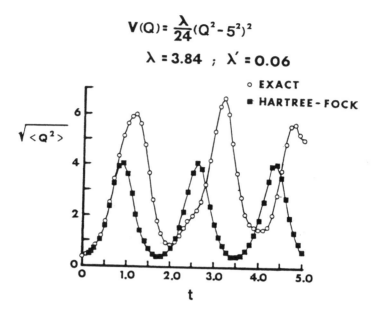

$$V(Q) = \frac{\lambda}{24}(Q^2 - 5^2)^2$$

$$\lambda = 3.84 \; ; \; \lambda' = 0.06$$

○ EXACT
■ HARTREE-FOCK

Figure 1a

$$V(Q) = \frac{\lambda}{24}(Q^2 - 5^2)^2$$

$$\lambda = 0.01 \; ; \; \lambda' = 1.06$$

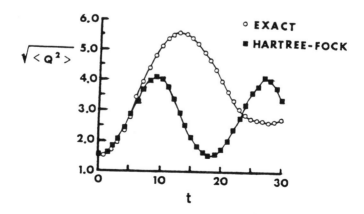

○ EXACT
■ HARTREE-FOCK

Figure 1b

$$V(Q) = \frac{\lambda}{24}(Q^2 - 5^2)^2$$

$$\lambda = 3.84 \; ; \; \lambda' = 0.06$$

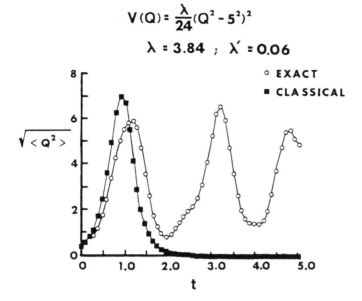

Figure 2

$Q_{\rm cl} < 4$ (the first oscillation) the two are quite the same except for a shift of the origin. We have already seen in Fig. 1 that the HF approximation is excellent until $\sqrt{G}$ reaches $\sqrt{\frac{2}{3}}\,a \approx 4.2$, and this implies that the classical behavior of $\sqrt{G}$ also appears in the HF approximation.

iii) The classical behavior in Eq. (3.3) which is found in the upside-down harmonic oscillator is tested: We now find that

$$-i\frac{\partial \psi_V}{\partial Q} = \left(\frac{i}{2}G^{-1} + 2\Pi\right) Q\psi_V \approx 2\Pi Q\psi_V \qquad (3.12)$$

and the ratio $2\Pi Q/p_{\rm cl}(Q)$, where $p_{\rm cl}(Q) = \sqrt{2\left(E - V(Q)\right)} = \sqrt{\frac{\lambda}{12}\left(a^2 - \frac{Q^2}{2}\right)}\,Q$, differs from unity by at most 20% for $1.5 < \sqrt{G} < 4$ in the HF approximation.

iv) The question whether the commutator $[Q, P]$ is negligible is studied both in the HF approximation and in the exact calculation: In the HF approximation,

$$\langle QP \rangle_{\rm HF} = \frac{i}{2} + 2G\Pi \qquad (3.13)$$

and we find that $G\Pi > 5$ for $1.75 < \sqrt{G} < 4$. In the exact calculation, the real part of $\langle QP \rangle_{\rm exact} > 5$ for even larger range $i.e.$ $1.2 < \sqrt{\langle Q^2 \rangle} < 5.9$.

Hence, in our variational time-dependent HF approach for small dimensionless coupling constant, $\lambda' \ll 1$ (or $\lambda > 1$) the late time behavior of the system is approximately described by classical physics with a classical probability distribution function,

$$f(Q, P, t) = |\psi_V(Q, t)|^2 \, \delta(P - p_{cl}) \quad . \tag{3.14}$$

For a large dimensionless coupling constant, $\lambda' = 1.06$ (or $\lambda = 0.01$) we find that classical behavior does not appear in the HF approximation nor in the exact simulation (see Fig. 1b).

IV. TIME-DEPENDENT VARIATIONAL APPROXIMATION FOR QUANTUM FIELD THEORY

Finally, I shall sketch how to generalize the approximation to a quantum field theory for a scalar field in a flat Robertson-Walker background metric

$$ds^2 = dt^2 - R^2(t)dx^2 \tag{4.1}$$

which may be used for studying inflation. In the functional Schrödinger picture, an abstract quantum mechanical state $|\psi(t)\rangle$ is replaced by a wave functional $\Psi(\phi, t)$, which is a functional of a c-number field $\phi(x)$ at a fixed time:

$$|\psi(t)\rangle \longrightarrow \Psi(\phi, t) \tag{4.2}$$

For the time-dependent HF approximation, we take a Gaussian trial wave function which is the generalization of Eq. (2.2)

$$\Psi_V(\phi, t) = N \exp\left\{ +\frac{i}{\hbar} \int_x \hat{\pi}(x, t) \left(\phi(x) - \hat{\phi}(x, t) \right) \right\} \exp -W$$

$$W \equiv \int_{x,y} \left(\phi(x) - \hat{\phi}(x, t) \right) \left[\frac{1}{4\hbar} G^{-1}(x, y, t) \right.$$

$$\left. - i\frac{1}{\hbar}\Sigma(x, y, t) \right] \left(\phi(y) - \hat{\phi}(y, t) \right) \tag{4.3}$$

where N is the normalization factor. Ψ_V is Gaussian centered at $\hat{\phi}(x, t)$ for each space-time point with width given by G. The conjugate momentum of $\hat{\phi}$ is $\hat{\pi}$ and Σ plays the role of the conjugate momentum of G. The variational parameters are ϕ, $\hat{\pi}$, G and Σ.

The Hamiltonian of a scalar field in the metric given in Eq. (4.1) is

$$H = \frac{1}{2} \int_x R^3 \left[-R^{-6} \frac{\delta^2}{\delta\phi^2} + R^{-2}(\nabla\phi)^2 + V(\phi) \right] \quad . \tag{4.4}$$

Then, the effective action Γ_V in this approximation is again given by Eq. (2.1), with $\psi(t)$ replaced by Eq. (4.3). The derivation of four variational equations are obtained straightforwardly by varying Γ_V against $\hat{\phi}$, $\hat{\pi}$, G and Σ.

This approximation allows us to study the onset of inflation for general classes of non-equilibrium initial conditions of a scalar field configuration. In particular, one can consider initial conditions where $\hat{\phi}(x, t) = \langle \phi(x) \rangle$ are randomly chosen in each initial horizon volume and/or with random initial momenta.[11] Furthermore, as discussed in section III, one can study the late time behavior of the scalar field. The numerical calculation for this problem is in progress.

REFERENCES

1. A.Guth, *Phys. Rev. D* **23**, 347 (1981).

2. A. Linde, *Phys. Lett.* **108B**, 389 (1982); A. Albrecht and P. Steinhardt, *Phys. Rev. Lett.* **48**, 1220 (1982).

3. A. Guth and S.-Y. Pi, *Phys. Rev. Lett.* **49**, 1110 (1982); S. Hawking, *Phys. Lett.* **115B**, 175 (1982); A. Starobinsky, *Phys. Lett.* **117B**, 175 (1982); J. Bardeen, P. Steinhardt and M. Turner, *Phys. Rev. D* **28**, 679 (1983).

4. G. Mazenko, W. Unruh and R. Wald, *Phys. Rev. D* **31**, 273 (1985); G. Mazenko, *Phys. Rev. Lett.* **54**, 2163 (1985) and University of Chicago preprint (1986); C. Coughlan and G. Ross, *Phys. Lett.* **157B**, 151 (1985); A. Albrecht, R. Brandenberger and R. Matzner, to be published.

5. A. Guth and S.-Y. Pi, *Phys. Rev. D* **32**, 1899 (1984).

6. F. Cooper, S.-Y. Pi and P. Stancioff, to be published in *Phys. Rev. D*.

7. R. Jackiw and A. Kerman, *Phys. Lett.* **A71**, 158 (1979).

8. P. A. M. Dirac, *Proc. Camb. Phil. Soc.* **26**, 376 (1930).

9. S. Epstein, *The Variational Method in Quantum Chemistry* (Academic Press, New York, 1976).

10. P. Bonche, S. Koonin and J. W. Negele, *Phys. Rev. C* **13**, 1226 (1976); A. Kerman and S. Koonin, *Ann. Phys.* (N.Y.) **100**, 332 (1976).

11. Albrecht *et al.*, in Ref. [4].

COSMOLOGICAL ANALYSIS OF R_p-BREAKING

Pierre Salati

LAPP, Chemin de Bellevue, BP. 909, F-74019 Annecy-le-Vieux Cedex
and Université de Savoie, F-73000 Chambéry

ABSTRACT

The breaking of R-parity allows the lightest supersymmetric
particle (LSP) to decay, and we study the cosmological
bounds on its mass and lifetime. These bounds can be translated into a lower limit on the neutrino mass when R-parity
is broken by a nonzero scalar neutrino v.e.v., and into
lower limits on the symmetry breaking parameters when R-
parity is broken by explicit couplings in the Lagrangian.

1. WHAT IS R-PARITY ?

Supersymmetric theories have been largely investigated during the
last decade. They predict, apart from the ordinary particles, a
spectrum of new particles. These supersymmetric particles differ from
the known particles by their R-parity. This parity is a Z-2 symmetry
and is defined as

$$R_p = (-1)^{2S+L+3B} \qquad (1.1)$$

where S, L and B are respectively the spin, the lepton number and the
baryon number. Supersymmetry associates an even-R_p number with the
ordinary particles and an odd-R_p number with their susy partners.

One generally builds models where the R-parity remains unbroken
after supersymmetry breaking, because breaking R-parity almost

inevitably leads to the violation of lepton or baryon number conservation, and these are supposed to be good symmetries at low energy (i.e. around 1 GeV) since no experimental evidence of such violations exists. However, L. Hall and M. Suzuki[1] concluded that particle physics cannot exclude R_p breaking, provided this breaking is quite weak. Therefore, the lightest supersymmetric particle (LSP), which in most supersymmetric models is the photino $\tilde{\gamma}$, can decay into ordinary particles but will be very long lived. Such a particle can play havoc in the early Universe. In this work[2], A. Bouquet and myself derive bounds on the mass and the lifetime of the photino by requiring the photino decay not to disturb the known evolution of the Universe.

2. TWO EXAMPLES OF R_p-BREAKING MECHANISMS

In order to disentangle the effects of the various R_p-breaking mechanisms, we will consider them separately.

2.1 The Spontaneous Breaking of the Lepton Number L

We consider the minimal supergravity-induced model. The relevant part of the potential of this theory is simply:

$$V = \frac{1}{8} (g_1^2 + g_2^2)(H^2 - G^2 + N^2)^2 + m_H^2 H^2 + m_G^2 G^2 + m_N^2 N^2 + 2\rho HG \qquad (2.1)$$

where H and G are the two neutral Higgs scalars, and N denotes one of the scalar neutrinos. The minimisation of V leads to the solutions :

i) N = 0. This corresponds to the usual R-parity preserving minimum;

ii) N $\neq$ 0. This solution leads to the spontaneous violation of L and, therefore, of R_p. However, this non-zero VEV for N is quite difficult to obtain[3].

A precise relation between the parameters of the potential V must be verified:

$$\rho^2 = (m_H^2 - m_N^2)(m_G^2 + m_N^2) \qquad (2.2)$$

Moreover, to obtain N^2 positive requires, at least, that:

$$m_H^2 > 2m_N^2 + m_G^2 \qquad (2.3)$$

In order to satisfy this bound (2.3), a considerable renormalisation of the previous parameters is needed between the GUT scale and the Weinberg-Salam breaking scale. This is only possible if the top quark is heavier than 70 GeV and the sleptons are lighter than 65 GeV.

The non-zero v.e.v. of the scalar neutrino N leads to the mixing of the photino with the corresponding neutrino and to the relation among their mass:

$$M_\nu \cdot M_{\tilde\gamma} = 0(<N>^2) \tag{2.4}$$

As the photino couples to the Z^o and the W through its mixing with the neutrino, it decays into a neutrino and a fermion-antifermion pair very much like a heavy neutrino. Its lifetime is given by

$$\tau = 1 \text{ second} \cdot \left(\frac{100 \text{ keV}}{<N>}\right)^2 \left(\frac{100 \text{ MeV}}{M_{\tilde\gamma}}\right)^3 \tag{2.5}$$

Finally, a non-zero sneutrino VEV leads to the existence of an axion like particle J, defined as:

$$J = -N + \frac{<N>}{<H>^2 + <G>^2} \left[<H>H - <G>G \right] \tag{2.6}$$

This particle is massless and couples to the electron through:

$$L_{eeJ} = \frac{m}{\sigma} \frac{<N>}{\sigma} eeJ \tag{2.7}$$

where σ is the Weinberg-Salam breaking scale $(\sigma^2 = <H>^2 + <G>^2)$.

2.2 The Explicit Breaking of R_p

The SU(2) doublets of superfields:

$$H = \begin{pmatrix} h^o \\ h^- \end{pmatrix} \quad : \quad \text{Higgs superfields}$$

and

$$L = \begin{pmatrix} \nu \\ e^- \end{pmatrix} \quad : \quad \text{Lepton superfields} \tag{2.8}$$

have a similar nature in supersymmetry since they carry out the same quantum numbers. The usual superpotential of the traditional supersymmetric extension of the Weinberg-Salam model

$$f_{traditional} = \lambda_U \; GQU + \lambda_D \; HQD + \lambda_E \; HLE + m \; HG \qquad (2.9)$$

may be completed by adding an explicit lepton number breaking part:

$$f_{\substack{lepton \; number \\ violation}} = \lambda_{LLE} \; LLE + \lambda_{LQD} \; LQD + m_{LG} \; LG \qquad (2.10)$$

or a baryon number breaking term:

$$f_{\substack{baryon \; number \\ violation}} = \lambda_{UDD} \; UDD \qquad (2.11)$$

The explicit violation of the lepton number leads to the instability of the photino with a lifetime given by:

$$\tau_{\tilde{\gamma} \to \gamma + \nu} = 10^{-10} \text{second} \left(\frac{M_{sfermion}}{100 \text{ GeV}} \right)^4 \left(\frac{1 \text{ GeV}}{M_{\tilde{\gamma}}} \right)^3 \frac{1}{\lambda^2_{\substack{LQD \; or \\ LLE}}} \qquad (2.12)$$

3. COSMOLOGICAL ANALYSIS

We analyse now the influence of an unstable photino on the first stages of Big-Bang. If the photino is light:

$$M_{\tilde{\gamma}} < 4 \text{ MeV} \qquad (3.1)$$

it thermally decouples from its surroundings when $t \sim 0.1$ second. The photinos are still ultrarelativistic when they freeze out, and their abundance is large:

$$\frac{n_{\tilde{\gamma}}}{T^3_{\nu}} = 15 \text{ cm}^{-3} \; {}^0K^{-3} \qquad (3.2)$$

On the contrary, if $M_{\tilde{\gamma}} > 4$ MeV, photinos annihilate among themselves as soon as the temperature underreaches $M_{\tilde{\gamma}}$. At a temperature T_q, significantly larger than T_f, the annihilation rate becomes less than the expansion rate and photinos are quenched. The annihilation stops. If photinos are stable, their density per volume that expands with the expanding universe remains constant. Figure 1 shows the present mass

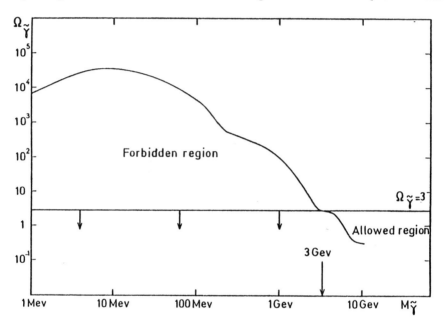

Fig. 1 : Relation between the cosmological parameter $\Omega_{\tilde{\gamma}} = \rho_{\tilde{\gamma}}/\rho_c$ (where $\rho_{\tilde{\gamma}}$ is the photino density) and the mass of the photino $M_{\tilde{\gamma}}$ for <u>stable photinos</u>. $H^0 = 100$ km/s/Mpc leads to $\rho_c = 2 \; 10^{-29} g.cm^{-3}$.

density $\Omega_{\tilde{\gamma}}$ of stable photinos as a function of the photino mass $M_{\tilde{\gamma}}$

$$\Omega_{\tilde{\gamma}} = \frac{\rho_{\tilde{\gamma}}}{2 \; 10^{-29} \; g.cm^{-3}} \qquad (3.3)$$

3.1 τ/M Limits

Figure 2 is the mass to lifetime plot of the photino. Six different bounds on these parameters can be set:

a) If the photino decays into ν_e neutrinos, there must exist today

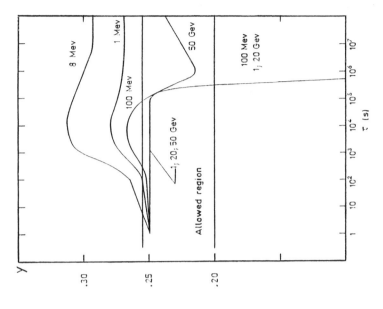

Fig. 3 : Variations with the photino lifetime $\tau_{\tilde{\gamma}}$ of the ^{4}He abundance by mass (Y) for different values of its mass $M_{\tilde{\gamma}}$.

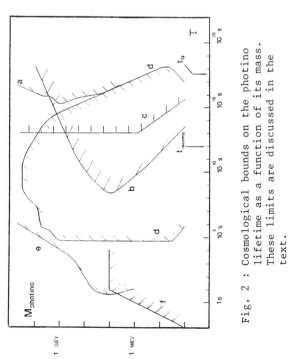

Fig. 2 : Cosmological bounds on the photino lifetime as a function of its mass. These limits are discussed in the text.

a cosmological background of ν_e. This background is susceptible to be detected by the Davis experiment of transmutation of Chlorine into Argon through the reaction:

$$^{37}Cl + \nu_e \rightarrow ^{37}Ar + e^- \qquad (3.4)$$

Since this ν_e background cannot exceed 6 Solar Neutrino Units, we derive limit a of figure 2.

b) The region to the right of curbe b of figure 2 is excluded. If the photino decays in this region of the plot, its decay products contribute too much to the present energy density of the Universe which cannot exceed $2 \cdot 10^{-29}$ g/cm^3.

c) If photinos release photons:

$$\tilde{\gamma} \rightarrow \gamma + X \qquad (3.5)$$

they will lead, through their decays, to an extra cosmic background ϕ_γ (photino) of gamma rays. An extra-galactic diffuse gamma ray background ϕ_γ (measured) was detected by several satellites, in particular SAS-2[4]. By imposing

$$\phi_\gamma \text{ (photino)} < \phi_\gamma \text{ (measured)} \qquad (3.6)$$

we obtained limit-c.

d) Any release of heat into the photon sea, between 10^5 seconds and the recombination time (i.e. 10^{12} seconds) leads to the increase of the entropy per baryon and to the distortion of the energy spectrum of the microwave background radiation (MBR). As the chemical potential μ of the MBR is constrained by observations:

$$\mu < 5 \cdot 10^{-3} \qquad (3.7)$$

the electromagnetic decays of the photino have to occur outside the forbidden region d of plot 2.

Primordial nucleosynthesis. Big-Bang nucleosynthesis provides us
th a nice explanation of the cosmic abundance of helium 4. As it
ands, it is the only sensible way to understand the observed contribu-
on Y of helium to the baryonic mass of the Universe (between 20%
d 25%). Since the theory nicely fits the observation, it provides
ringent constraints on any new particle that would alter the standard
enario of primordial nucleosynthesis. Indeed, the possible existence
the photinos affects primordial nucleosynthesis in three different
ys:

) By their presence, they increase the total energy density of the
Universe, and therefore accelerate its expansion. The freezing of
the neutron/proton ratio occurs earlier, and this ratio is then
bigger than in the standard case. This results in a significant
increase of Y.

i) The electromagnetic decay products of the photinos interact with
the primordial plasma that fills up the Universe. This leads to a
reheating of the photon gas, and to the increase of the entropy
per baryon. The subsequent drop of the baryon to photon ratio η
largely upsets the synthesis of the light elements D, ^{3}He, ^{4}He...
This ratio η not only may change by an order of magnitude, but
continuously evolves with time. The light nuclei abundances cru-
cially depend on η, the variation of which must carefully be
taken into account. The reheating of the primeval plasma also
modifies the time to photon temperature profile. The decrease of
the temperature is slowed down, owing to a modification in the
kinetics of the nuclear reactions. This leads to a change in the
cosmological abundances of the light elements.

ii) Last, but not least, the photodisruption of nuclei. If the photinos decay during or after the nucleosynthesis, the decay photons
would photodissociate D, ^{3}He, ^{4}He ... This process depends on
the photon energy which must be larger than the thresholds of the
relevant photodestruction reactions. A crucial role is played by
the electromagnetic showers that are induced by radiative decays
of the photinos.

A detailed analysis of these three effects was recently completed[5]. The authors showed that the standard results of nucleosynthesis may be drastically changed if a substantial part of the elements are photo-destroyed into hydrogen. Photodissociation may be very efficient since in some cases $(M_{photino} > 40$ MeV and $\tau > 10^5$ seconds), the Universe ends filled up with protons only once the photinos have decayed. As an illustration, Figure 3 displays the variations with the photino life-time $\tau_{\tilde{\gamma}}$ of the ^{4}He abundance by mass (Y) for different values of its mass $M_{\tilde{\gamma}}$. This analysis leads to the bound e of fig. 2.

f) The study by Falk and Schramm[6] of the dynamics of supernovae explosions sets the constraint f of plot 2 on the photino lifetime:

$$\tau < 10 \text{ seconds} \cdot \frac{M_{\tilde{\gamma}}}{10 \text{ MeV}} \tag{3.8}$$

3.2 The Axion Like Bound

The J axion like particle, if it exists, must couple very weakly to the electron. Otherwise, J is thermally produced in stars through the reaction:

$$e\gamma \rightarrow eJ \tag{3.9}$$

By freely escaping and carrying out the internal heat, J overcools stars. In particular, J supercools the core inside red giant stars. In order to avoid the premature death of these stars, <N> has to be less than 9 keV[7].

4. THE RESULTS

4.1. Spontaneous L Breaking

Figure 4 is a $M_\nu - M_{\tilde{\gamma}}$ plot where the cosmological bound derived from this analysis, the axion like bound and the Sarkar-Cooper limit (i.e. the mass of the neutrino is less than 100 eV[8]) are displayed. Combining these three limits leads to:

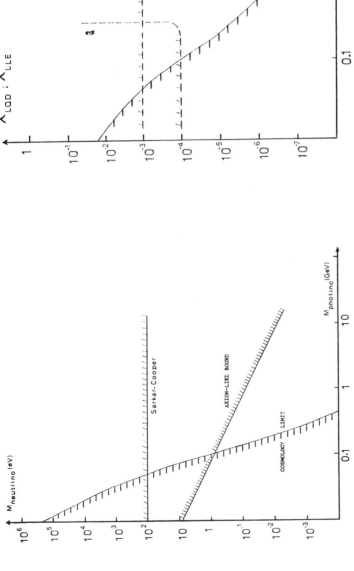

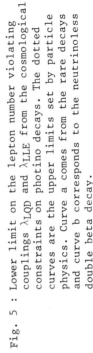

Fig. 5 : Lower limit on the lepton number violating couplings λ_{LQD} and λ_{LLE} from the cosmological constraints on photino decays. The dotted curves are the upper limits set by particle physics. Curve a comes from the rare decays and curve b corresponds to the neutrinoless double beta decay.

Fig. 4 : Bounds on the neutrino mass. The heavy line shows the lower limit set by constraints on photino decays, and must be contrasted with the upper cosmological bounds established by Sarkar and Cooper[8] and by Dearborn, Schramm and Steigman[7]).

$$M_\nu < 0.3 \text{ eV}$$
$$M_{\tilde\gamma} > 200 \text{ MeV}$$
(4.1)

A very light photino $(M_{\tilde\gamma} < 100 \text{ eV})$ is also possible provided that, in this case, its lifetime exceeds the age of the Universe.

4.2 Explicit R_p-Breaking

We start the same game in Figure 5. The non observation of rare decays such as:

$$\pi^+ \to e^+ \tilde\gamma \; ; \quad K^+ \to e^+ \tilde\gamma \; ; \quad K^+ \to \mu^+ \tilde\gamma$$
(4.2)

provides the constraint a. The neutrinoless double beta decay bounds translate into curve b. By combining these particle physics data with t cosmological limit, we extract from this plot $\text{Log}(\lambda_{QDL})/M_{\tilde\gamma}$ the limit:
$$\text{LLE}$$

$$M_{\tilde\gamma} > 100 \text{ MeV}$$
(4.3)

Once more, we cannot exclude a light photino if $\tau > t_u$.

ACKNOWLEDGEMENTS

I would like to thank the organizing Committe for financial suppor

REFERENCES

1. Hall L.J. and Suzuki M., Nucl. Phys. **B231**, 419 (1984).

2. Bouquet A. and Salati P., PAR-LPTHE 86-09 / LAPP-TH-157, to be published in Nuclear Physics.

3. Gato B., Leon J., Perez-Mercader J. and Quiros M., Nucl. Phys. **B260**, 203 (1985).

4. Trombka J.I. and Fichtel C.E., Phys. Rep. **C97**, 173 (1983).

5. Audouze J., Delbourgo-Salvador P. and Salati P., IAP and LAPP-TH-148 preprint April 1986, to be published in Astronomy and Astrophysics.

6. Falk S.W. and Schramm D.N., Phys. Lett. **79B**, 511 (1978).

7. Dearborn D.S.P. Schramm D.N. and Steigman G., Phys. Rev. Lett. **56**, 26 (1986).

8. Sarkar S. and Cooper A.M., Phys. Lett. **148B**, 347 (1984).

Strings, the Peebles Screed, and Large Scale Structure

David N. Schramm
The University of Chicago and Fermilab

Abstract
The combined problems of large scale structure, the need for non-baryonic dark matter if $\Omega = 1$, and the need to make galaxies early in the history of the universe seem to be placing severe constraints on cosmological models. In addition, it is shown that the bulk of the baryonic matter is also dark and must be accounted for as well. The nucleosynthesis arguments are now strongly supported by high energy collider experiments as well as astronomical abundance data. The arguments for dark matter are reviewed and it is shown that observational dynamical arguments and nucleosynthesis are all still consistent at $\Omega \sim 0.1$. However, the inflation paradigm requires $\Omega = 1$, thus, the need for non-baryonic dark matter. A non-zero cosmological constant is argued to be an inappropriate solution. Dark matter candidates fall into two categories, hot (neutrino-like) and cold (axion or massive photino-like). New observations of large scale structure in the universe (voids, foam, and large scale velocity fields) seem to be most easily understood if the dominant matter of the universe is in the form of low mass ($9eV \leq m_\nu \leq 35eV$) neutrinos. Cold dark matter, even with biasing, seems unable to duplicate the combination of these observations (of particular significance here are the large velocity fields, if real). However, galaxy formation is difficult with hot matter. The potentially fatal problems of galaxy formation with neutrinos may be remedied by combining them with either cosmic strings or explosive galaxy formation. The former naturally gives the scale-free correlation function for galaxies, clusters, and superclusters. The latter requires fine tuning and percolation to get the large scales and the scale-free correlation function. However, combining hot matter and strings reduces the ability of the hot matter to give some of the large scale features and still yield $\Omega = 1$. Questions to be examined are raised.

1. Introduction

The major confrontation of early universe studies with the "real" universe now focuses on the problems of galaxy formation, dark matter, and the generation of large scale structure. The observable aspects of these problems came into being shortly after recombination; however, the condition of the universe as it approaches recombination are determined by events taking place much earlier, when nuclear and particle physics effects dominated. Since the recombination epoch is the limiting epoch for direct observations, it is only natural that this epoch serve as the interface between early universe cosmologists and astronomers.

The problems are to produce initial conditions and types of matter which will yield the observable universe, the large scale structure. In particular, the observable universe now appears to have large scale structure on scales of $\sim 40Mpc$ that looks like foam or at least intersecting sheets and filaments with large voids[1,2,3]. In addition, there appear to be large, coherent motions of 40 Mpc clumps with velocities of $\sim 600km/sec$[4]. To this very large scale structure must be added the apparent fact that clusters of galaxies cluster with each other more strongly than galaxies cluster[5], or to use the analysis of Szalay and Schramm[6], the clusters and galaxies appear to cluster in a scale-free manner as if laid out in some fractal pattern.

2. The Dynamical Arguments

To these large scale observations must be added the dynamical measurements of mass and the so-called dark matter problem. In particular, the dynamics of the visible parts of galaxies imply an Ω of ≤ 0.01 (where $\Omega \equiv \frac{\rho}{\rho_{crit}}$ is the critical density of the universe). However, when galaxies interact with other galaxies in binary pairs or in small groups, they interact with ~ 10 times as much mass, implying an $\Omega \sim 0.1$. When galaxies interact with one another in large clusters they interact with possibly even more mass, implying $\Omega \sim 0.1$ to 0.3. (*No well studied system gives anything near $\Omega = 1$.*)

3. Big Bang Nucleosynthesis

To the dynamical arguments we can add the arguments from Big Bang nucleosynthesis (Yang et al.) which show that observed abundances are consistent only if $\Omega_b \sim 0.1$ (where $\Omega_b \equiv \frac{\rho_b}{\rho_{crit}}$ and ρ_b is the density of baryons).

Thus as Gott et al.[7] pointed out over ten years ago, direct astronomical evidence points towards $\Omega \sim 0.1$ with the dark halos being baryonic and no need for exotic stuff. In particular, it should be noted that the lower bound on Ω_b is $\Omega_b \geq 0.03$[8]. Since this is > 0.01, it implies that the bulk of the baryons are dark. (Note that because of this point, dark halos for dwarf spheroidal galaxies are no problem since they can be baryonic.) Also, it is important to remember that nucleosynthesis contrains $\Omega_b < 0.15$. (This is lower than the 0.19 from Yang et al.[9] due to better current upper limits on the microwave background temperature.) Thus, if $\Omega \sim 1$, the bulk of the universe would be non-baryonic *and* could not cluster with the light emitting galaxies and clusters.

The nucleosynthesis arguments are gaining even greater credence now that their prediction[9,10], that the total number of neutrino types (generations) is small (three or at most four) is being verified by collider experiments[11] with current experimental limits at < 5. From particle physics theory alone any number of generations might be possible. The preliminary verification of the cosmological prediction is the first time that cosmology·has made a prediction which has been verified by a high energy accelerator experiment.

4. Baryonic Halos?

Can halos of galaxies and dwarf spheriodals really be baryonic? While the coincidence of $\Omega_b \sim 0.1$ and $\Omega_{dynamic} \sim 0.1$ is suggestive, it is certainly not compulsory. Different forms of dark matter can mix with baryons in different ways depending on the mechanism of galaxy formation.

With cold dark matter the halos must be a mixture of $\sim 90\%$ cold matter and 10% baryons whereas in hot matter models the halo mixture depends on the galaxy formation scenario.

If the halos do contain significant baryonic materials, what form can it be? Hegyi and Olive and Schramm have argued that most baryonic things do not work. However, they leave two very important loopholes:

1. Black holes left from an early generation of massive stars with the bulk of the stellar material falling into the hole and not producing excess heavy elements. Such black holes are contrained by Big Bang nucleosynthesis baryon limits since they were baryons then (so they count as baryonic material).

2. Low mass objects too dim to be seen in telescope searches. Jupiter-like clumps or even 0.01 $M_\odot$ stars would work. In order for the abundance of such objects to be sufficient, the abundance spectrum for these objects would probably be above the low mass extrapolation of the Salpeter initial mass function. However, that function is strictly empirical and there could certainly be a low mass excess if the initial stellar generation with pure H and He, but more objects low than currently occurs with heavy elements present. (Option 1., of course, requires exactly the opposite behavior for the early stellar mass function.)

5. The Flatness Arguments

If everything agrees so well with $\Omega \sim 0.1$, why do people continue to think $\Omega = 1$? The only astrophysical evidence for large Ω is clearly weak at the present time. It consists of the following:

1. With Gaussian adiabatic initial density fluctuations of the type described by Zel'dovich and expected from simple inflation models, it is impossible to make galaxies rapidly enough when constrained by limits on microwave background anisotropies unless $\Omega > 0.2$[12,13].

2. The velocity field of IRAS galaxies on scales of ~ 200Mpc implies a virial mass on these large scales of $\Omega \sim 1$[14].

3. The density of galaxy counts versus redshift is optimally consistent with $\Omega = 1$ geometry[15].

The first of these is clearly removable if galaxies form by something other than Gaussian adiabatic fluctuations with a Zel'dovich spectrum. In particular, string models which are also derivable from grand unified gauge models do not yield such a stringent requirement on Ω, nor do, for that matter, models where galaxy formation is stimulated by early explosions[16].

The second argument has the problem that a reliable way to determine distances to IRAS galaxies has not been established and a complete redshift survey of IRAS galaxies remains to be done. In addition, IRAS counts may have a significant north-south bias due to induced instrumental variations in sensitivity of the satellite in the northern and southern hemispheres.

The third argument, while potentially the strongest, still requires a more detailed analysis of galactic evolution effects and normalization of distant galaxy counts to nearby where different techniques are used.

Thus, while suggestive, these arguments do not yet establish $\Omega = 1$. However, there is a Copernican-like argument which is sufficiently powerful that most theoretical physicist believe $\Omega = 1$. The argument was best articulated by Dicke and Peebles and later provided Guth with a strong motivation for inflation which gave a physical mechanism for yielding the desired Ω. The argument, simply stated, is that Ω is a time changing quantity going to $\Omega < 1$ and to ∞ if $\Omega > 1$, and only remaining constant if $\Omega = 1$. The timescale of change is the expansion rate of the universe. Thus, the only long-lived values are 0, 1, and ∞. Since we are here, Ω is neither 0 nor ∞. The only other long-lived value is 1. To have any finite value below unity today would require that we live at a very special time, the early epoch in cosmic time when Ω was not 1 or 0. Such a value would require the extraordinary fine tuning at the Planck time of ~ 60 decimal places, or at least 17 decimal places at the time of Big Bang nucleosynthesis. Thus, unless we live at a special time and some unknown mechanism tunes Ω to exactly the right amount to fantastic accuracy, Ω is probably unity.

Since any early deSitter phase for the universe produces a flat universe ($\Omega = 1$ if the cosmological constant $\Lambda = 0$) and since inflation means an early deSitter phase, and since most scalar fields yield inflation, it is reasonable to believe $\Omega = 1$. While many have recently focused on the problems many models of inflation have been producing, the right sized initial fluctuations[17] any inflation model which solves the horizon problem, getting a nearly constant background temperature, will also solve the flatness problem.

6. The Cosmological Constant

Some astrophysicists (who shall remain nameless) have focused on the formal mathematical loophole that flatness can also be obtained with a non-zero $\dot{\Lambda}$ and $\Omega < 1$. However, such a solution is missing the philosophical motivation (like killing for pacifism). If today we have $\Omega \sim 0.1$ and non-zero Λ yields flatness, that is an epoch- dependent solution since the contribution of Ω and Λ vary differently with epoch. Such a solution would imply that we live at the only epoch where Λ and Ω contributions to curvature are comparable, again requiring amazing fine tuning (tuning Λ to ≥ 120 decimal places). Unfortunately we don't as yet have a nice physically motivated mechanism like inflation to set $\Lambda = 0$, but if we buy the philosophy, I believe we should also assume Λ is negligible. Of course both arguments are philosophical (or theological) rather than based on physical observation, but the Copernical principal of us not being special has held up well for several hundred years.

7. Dark Matter and Galaxy and Structure Formation

As mentioned before, if Ω is 1, then we need non-baryonic dark matter. Such matter has been classified as either hot (neutrino-like with high velocities just prior to the epoch of matter-radiation equality) or cold (low velocities prior to matter-radiation equality).

Initially, hot, low mass, neutrinos were quite popular as candidates for solving the cosmological dark matter problem, since they were the least exotic of the non-baryonic options, and they naturally clustered only on large scales where the dark matter was needed, rather than on the small scales where the contribution of dark matter was known to be minimal[18]. They received a major boost with the preliminary reports of measured mass[19] for ν_e (although probably only the most massive ν is cosmologically important, and that might well be ν_τ (or a nucleosynthesis-allowed 4th generation) which could still have a $\sim 10eV$ mass, even if $m_{\nu_e} \ll 1eV$). Also, they gained strength when it was shown[3] that the neutrino Jean's mass was

$$M_J \sim \frac{3 \times 10^{18} M_\odot}{m_\nu^2 (eV)} \text{ or } \lambda_J \sim \frac{1300 Mpc}{m_\nu (eV)}$$

which for $m_\nu \sim 30eV$ yielded $M \sim 3 \times 10^{15}M_\odot$, and $\lambda \sim 40Mpc$, the mass and scale of large clusters.

Unfortunately, massive neutrinos fell into disrepute as dark matter when it was emphasized[20] that in the standard adiabatic model of galaxy formation with a random phase, Zel'dovich fluctuation spectrum of the type expected by inflation, and with $\delta T/T$ constrained by microwave observations, galaxies did not form until redshift $z \lesssim 1$. This occurred because the initially formed pancakes with mass M_J took a while to fragment down to galaxy size. This contradicted the observations which showed that quasars existed back to $z \sim 3.5$. In addition, if baryons stay in gas form in the potential wells of the large ν pancakes, they light up in the x-rays beyond what is observed[21].

While some[22] have appealed to statistical tails, etc., to escape these conclusions, most cosmologists began abandoning neutrinos and adopting cold dark matter[23], which could enable rapid galaxy formation[24,25].

Cold matter also had its problems[26]. In the standard model, it would all cluster on small scales, and thus be measured by the dynamics of clusters, such as the Virgo infall. Since such measurements implied that $\Omega \sim 0.2 \pm 0.1$ on cluster scales, this meant that $\Omega_{cold} \lesssim 0.3$, and not unity. Remember that $\Omega \sim 0.1$, so observationally, non-baryonic dark matter is not required unless one wants an Ω of unity, so cold matter wasn't naturally solving one problem for which it was postulated. This constraint on cold matter could be escaped if it were also assumed that galaxy formation was biased[25,27] and did not occur everywhere. Thus, there could be many clumps of cold matter and baryons that did not shine for some ad hoc reason. Biasing ran into problems when it could not explain the observation[5] of a very large cluster–cluster correlation function, ξ_{cc}, relative to the galaxy–galaxy correlation function[26,27], ξ_{gg}. With biasing $\xi_{cc} \propto \xi_{gg}$ but in all models $\xi_{gg} < 0$ for a few 10's of Mpc, whereas ξ_{cc} was observed to be positive out to scales $\gtrsim 50Mpc$. Hardcore cold matter lovers had to argue that the ξ_{cc} data might be wrong, although no one has been able to disprove it.

A way out of the ξ_{cc} problem was proposed by Szalay and Schramm. There we noted that the correlation functions appear to be scale free, thus implying that large-scale structure is dominated by something other than random noise and gravity, say either percolated explosions or strings. In fact, the scale-free structure is characterized by a fractal of dimension $D \sim 1.2$, not too different from the $D \sim 1$ that naive string theory might yield. String calculations[28] of galaxy formation indeed found support for such a fractal process with the appropriate dimension being valid from galaxy to supercluster scales.

Thus, there were already strong hints that something was wrong with the previous, in vogue, picture of biasing and cold matter with random noise initial fluctuations. To this we now add the new observations of many large voids[1,2] of diameter $50h_{1/2}Mpc$ ($h_{1/2} \equiv H_0/50km/sec/Mpc$), with most galaxies distributed on the walls of the voids, and the observation[4] that our local 40 Mpc region of space is moving with a coherent velocity field of $\sim 600km/sec$ toward Hydra-Centaurus. While at least one large void (in Böotes) had been observed before[3], using a pencil beam approach, until the Harvard redshift[1] survey work, it was not known how ubiquitous voids were. In fact, the Harvard data shows that almost all galaxies are distributed along the "walls" of voids; galaxies and clusters are not randomly distributed, but fit onto a well-ordered pattern.

While the Harvard work only goes out to $\sim 100Mpc$, there is substantial evidence that this sort of pattern persists to redshifts $z \sim 1$ from the Koo and Kron survey[2]. A simple explanation for the peaks and valleys in the distribution of galaxies and quasars with redshift is that one is looking through filaments or shells with voids in between, once again demonstrating that galaxies and clusters are not laid out randomly on the sky, but follow a pattern.

While statistical fluctuations with cold matter might yield a few large voids as well as many small voids[21,25], it is difficult to get all of space filled with large voids and have galaxies appear only at the boundaries unless some special form of "biasing" is used. However, the real killing blow for the cold matter plus biasing scheme comes from the velocity field work. Even if the biasing could be selected so as to give ubiquitous large voids, the velocities of a $40Mpc$ region of galaxies would be relatively small and random, rather than large and coherent[29]. In fact, the more extreme the biasing used to get large voids, the lower the large scale velocities. Thus, it appears that the large-scale structure is telling us that we need something that gives us $\sim 40Mpc$ coherent patterns, and cold matter doesn't appear the way to go. (Unless, of course, the large scale velocity field work is in error. In other words, cold matter with gaussian Zel'dovich fluctuation requires *both* ξ_{cc} and the velocity to be completely wrong.

Since neutrinos naturally gave us patterns on this scale, maybe they should be reexamined. In addition, since the voids look rather spherical, and since explosions tend to produce spherical holes after a few expansion times even if the initial explosion is asymmetric, perhaps an explosive mechanism should be considered also. Since the Ostriker–Cowie[16] explosion mechanism by itself cannot yield such large voids, the only way it could work is via a high density network of explosions which percolated[25,30]. However, to get $\Omega = 1$ with an exploding scenario would still require non-baryonic matter that did not cluster with the light emitting stuff. In principal, this could be either neutrinos or cold matter but at least with neutrinos an $\sim 40Mpc$ scale might still be naturally imposed.

8. Neutrinos plus Strings or Explosions

Of course, in order for neutrinos to work as the dominant matter, some mechanism to rapidly form galaxies must be imposed both to enable galaxies to exist at $z \sim 5$, and to condense out the gas before it falls into the forming deep potential wells, and emits x-rays. Two ways that might achieve this rapid formation are either via the aforementioned explosion scheme within the collapsing ν-pancakes, or via cosmic strings[31] which would act as nucleation sites for galaxy formation. Since strings are not free-streamed away by the relativistic neutrinos[32], the galaxy scale fluctuations remain within the ν-pancakes. Notice that since neutrinos are not used by themselves simple arguments based on relating their primordial fluctuation spectrum to observed galaxy velocity and distribution features are not necessarily valid and must be reexamined in the more complete scenario.

It should be noted that even with strings as seeds so that cold matter can cluster in a scale-free way fitting ξ_{cc}, the large scale velocity fields for cold matter are small, and it is difficult to get $\Omega = 1$ while observing $\Omega_{cluster} \sim 0.2$. However, we have the additional problem that the strings might mess up the nice large scale neutrino features and background of ν's will still slow galaxy growth around the strings over how cold matter would form on the strings.

It is interesting that two surviving galaxy formation options, strings and explosions, involve the same two options that the scale-free cluster–cluster correlation function arguments point towards. Let us look at each of these scenarios in a little more detail and see if there might be ways of resolving whether either of them might actually be correct. Also, let us see what each requires for the physics of the early Universe.

Both of these scenarios seem to need hot matter if we want to solve the velocity field, $\Omega = 1$, and large scale problems. If $\Omega = 1$, as is necessary to avoid our living at a special epoch, and as agrees with the recent large-scale galaxy count arguments of Loh and Spillar[15] (but disagrees with the direct dynamical arguments on scales of clusters and smaller, and with the baryonic measurements from nucleosynthesis), then $m_\nu \lesssim 35eV$. Since with $\Omega = 1$ the age of the Universe $t_0 = \frac{2}{3H_0}$, and since globular clusters and nucleochronology require $t_0 \gtrsim 11 \times 10^9 yr$ (with a best fit of $t_0 \sim 15 \times 10^9 yr$) we must say that $H_0^{-1} \gtrsim 17 \times 10^9 yr$. Thus, $H_0 \lesssim 60km/sec/Mpc$, or $h_{1/2} \lesssim 1.2$. From the number of neutrinos and photons in the Universe, we know that the most massive neutrino is bounded by (see ref. 18 and references therein)

$$m_\nu \lesssim (25eV)\Omega h_{1/2}^2 \lesssim 35eV.$$

It is curious that the requirement that we want the neutrinos to give us the large-scale structure, $\lambda_J \sim 40Mpc$, or $M_J \sim 10^{16}M_\odot$, also gives us $m_\nu \sim 30eV$, a mass about what is necessary to get $\Omega \sim 1$. Also, we have a lower bound from the nucleosynthesis argument[26] that the number of neutrino species with $m_\nu \lesssim 10MeV$ is three or at most four. Since the sum of all neutrino masses cannot exceed the $35eV$ limit mentioned above, and since the lowest mass for the most massive one occurs when they are all equal, then if $N_\nu \leq 4$,

$$m_\nu \gtrsim 9eV.$$

The first scale to be able to condense and thus have their density grow will be the horizon scale when the neutrinos become non-relativistic, which is M_J. However, in the string option, loops of string will exist down to scales of galaxy size (scales smaller than galaxy size gravitationally radiate away[31]). So as the neutrinos become non-relativistic they can be trapped on smaller scales. The baryons will not be able to begin clustering until after recombination. However, the slow-moving baryons will rapidly fall on to the pre-existing loops of string plus neutrinos. Thus, galaxies will be able to form shortly after recombination, and well before $z \sim 1$.

9. Problems with Strings?

Unfortunately, just after matter domination the bulk of the neutrinos will still have relatively high velocities so their Jean's mass, while dropping, will not be low enough for most neutrinos to cluster on the galaxy size loops. Even after recombination the characteristics Jean's mass for the bulk of the neutrinos will still be much larger than galaxy size, so there will be a relatively smooth background of neutrinos which will slow the rate of growth of baryons falling onto the loops of string. Thus, strings plus neutrinos do not grow galaxies as rapidly as strings plus cold matter; however, strings definitely help the neutrino picture along. The quantitative question of whether the neutrino-string picture can form rapidly enough remains to be worked out in detail, since quick and dirty calculations indicate that the results are marginal[33]. With neutrinos, the dimensionless string tension 6μ needs to be higher than for strings with cold matter where $6\mu \sim 10^{-6}$. Unfortunately, it cannot be arbitrarily raised since high values ($\geq 10^{-5}$) cause problems in microwave anisotropy and in radiating too much energy at the time of nucleosynthesis, thus running into the equivalent of the neutrino country bound[34].

Also, it is not clear how the combination of ν's and strings deals with the very large scale structure. While strings by themselves give the scale- free correlation function out through the scales of Abell clusters[28], if neutrino pancaking is too strong, it could mess this up. On the other hand, string perturbations existing on scales smaller than ~ 40Mpc may prevent pancaking from occurring at all. Horizon length strings at matter-radiation equality will produce large scale adiabatic fluctuations that could induce pancake formation in the neutrinos, going non-linear at redshift $z \sim 1$. However, the strength of the fluctuations relative to the normal string fluctuations needs to be checked to see which, if any, dominates.

If they really do not go non-linear until $z \sim 1$, they might not mess up the more rapidly forming galaxy and cluster scale fluctuations, so the smaller scale correlation functions might be retained while the neutrino pancake collapse might induce the very large scale velocity field and pancakes, filaments, and voids. Obviously the whole combined picture needs to be examined in much greater detail to see if it really can retain the best features of both models, rather than the two components destroying each others better features.

Because the string picture looks like the current front runner, people have begun looking at it in far greater detail, to see if it really can yield the observable universe. In particular, Peebles has privately circulated a "screed", stating possible problems. At a workshop held at the Aspen Center for Physics, these problems were examined and possible ways out were found. Let us now summarize the Peebles problems and possible solutions.

Problems not previously mentioned:

1. Strings produce loops following a power spectrum $\sim M^{-5/2}$, whereas galaxies are observed from their light to have a much flatter spectrum, up to $\sim 10^{12} M_\odot$ and then exponential fall off. Thus, at first glance, it appears that strings give too many small *and* large galaxies if their spectrum is normalized to fit the L^* galaxies at $\sim 10^{12} M_\odot$.

2. Strings are small relative to their separation distances. Thus, collapse onto static strings appears unlikely to give large quadrupole moments, and thus tidal interactions will not produce the angular momentum observed in galaxies.

3. With strings as seeds, both cold and hot dark matter will cluster on small scales so that Ω measured for clusters should be a good estimate of Ω_{total} which would yield ~ 0.2, not 1. Biased suppression of galaxy formation with strings as seeds is evn more ad hoc than normal cold-matter biasing, so is not a convenient escape.

The possible solutions to these problems are:

1.1. Excess amounts of small strings forming galaxies can be supressed in a variety of ways.

    a. For larger 6μ, such as in the neutrino models, gravitational radiation eliminates the excess low mass loops.

    b. Vilenkin[35] has shown that global strings rather than gauge strings radiate Goldstone bosons in addition to gravitational radiation. Thus few mass global strings would also not be a problem.

    c. Strings do not radiate symmetrically. The differential radiation for small strings results in a rocket effect[36] which supresses their ability to acrete.

    d. More fragmentation of the small loops which form early could lower their abundance as the smaller are radiated away.

1.2. The excess amounts of large loops may be a more complex problem and more work needs to be done here. Possible solutions include:

    a. Finite velocity may affect accretion.

    b. Fragmentation of large loops will reduce their numbers.

    c. Big loops may yield CD galaxies at centers of clusters with velocity curves rising as $r^{1/4}$ rather than normal flat rotation curves.

2. Angular momentum may be formed by tidal interactions because accretion is not spherical but sausage-like, due to the finite velocity of loops. Distances moved are comparable to separations so quadrupole moments will be approximately large.

3. The solution to the Ω problem requires that somehow clusters don't sample a standard segment of the universe. One way to accomplish this would be if galaxies correlated more with clusters than randomly. Such could occur if large, cluster-producing strings fragment to produce smaller galaxy-producing strings, and the resultant small strings didn't get too far from the clusters. Clearly, this does occur to some degree; however, can it quantitatively yield a factor of three of more enhancement in Ω between its cluster measured value and the true value remains to be shown. The dynamical range of string simulations has not yet enabled such quantitative tests between small and large loops. Note that if galaxy strings are strongly correlated with clusters, then many regions in space will be without loops of strings, and so will not form galaxies even though they have baryons and either hot or cold dark matter.

Another possible problem is that, while the string scenario may naturally yield $D \sim 1$, it does not so naturally give $D = 1.2$. Fine tuning[39] of string parameters may enable such variation on the scale of the galaxy–galaxy correlation function, or some modification of the criteria for the formation of light-emitting regions around the strings may be necessary.

In this regard it should be remembered that because of possible systematic errors, not everyone agrees that 1.2 is significantly different from 1.0, even for the galaxy–galaxy correlation function, which is the best determined[38]. The uncertainties in the exponent of the cluster–cluster correlation functions are far larger, thus problems in trying to explain variations from $D = 1$ fractals are not serious at the present time. With strings there is the additional problem of tuning the primordial phase transition so as to inflate first, and then produce strings[39]. While not impossible, this is constraining.

10. Explosive Galaxy Formation

The second way to get neutrinos to work involves explosive galaxy formation. Here we need initial seeds to lead to condensations which produce massive baryonic objects which explode. As mentioned before, such a model does not naturally give us $40Mpc$ structure. If we use neutrinos then the seeds must be in a form which does not get free-streamed away by the relativistic neutrinos. Strings don't work well here because the string scales that might lead to rapidly evolving baryonic objects are radiated away gravitationally. Thus, the seeds must come in some other isothermal-like form. Perhaps the best option would be condensates from the quark–hadron transition, either planetary mass black holes[40] or Witten nuggets[41]. Both have formation problems[42] and the latter have survival problems[43] also. If such objects could form and survive, they do lead naturally[44] to very massive ($\sim 1000M_\odot$) baryonic objects which would explode on rapid timescales. Another option is cold dark matter clumps, in which case small strings work as seeds, but the large scale problems are aggravated.

194

The scale affected by explosions of single galaxy size[45] is at most a few Mpc; however, it has been shown[30] that at sufficiently high densities and high trigger rates, the explosions can percolate at least out to scales of a few 10's of Mpc. The fractal dimension of such percolated ensembles is quite sensitive to parameter assumptions and usually varies with scale, thus showing that it is not a true scale-free fractal. If it is made to fit the small scale (few Mpc) with $D \sim 1$ it is usually larger ($D \sim 2$) on scales of $\gtrsim 10Mpc$. Since, as mentioned above, the exponent of the cluster–cluster correlation function is not, at present, well determined, such models cannot be ruled out. With such explosions percolating within ν-pancakes, we might naturally have their pattern superimposed on the $\sim 40Mpc$ neutrino scale. In addition, although percolated explosions will initially be highly non-spherical, their shape will evolve towards sphericity with the smaller axes catching up in length to the largest one. In order for large-scale percolation to occur, several generations[21] of explosions must occur; however, cooling arguments and time to initial explosions, plus the need for condensed objects by $z \sim 4$ and the need to hide from present observers, the radiation produced by the explosions, severely restrict the possibility of such percolation and thus quite a bit of fine tuning is required to escape the constraints.

11. Conclusion

Thus, while we cannot explicitly rule out this latter case, unless some new physics can be developed to show how the fine-tuned parameters are natural for other reasons, we must lean towards the string option as the present frontrunner. Strings, of course, would have other observational consequences[32] like gravitational double lensing of distant objects and shifts in the 3° background across such a line of lenses, and a background of gravitational radiation from the evaporation of small-scale strings which might affect the millisecond pulsar. Thus, observations should eventually be able to confirm or deny this frontrunner. Table 1 gives a summary of current proposed models and their ability to solve the problems. Note that the location of dark baryons may eventually be detectable and a discriminator of models. No model is yet a clear winner. Some require more calculations to see if they can be made to work. Others require some key bit of observational data to be proven wrong.

In summary, we have come full circle and once again massive neutrinos are looking good. However, with them comes the need for galaxy and structure formation triggered by something other than random phase adiabatic fluctuations. The non-random phase fractal initial conditions such as produced by strings[46] or fractal generating explosions[16,30] seem to be the way to go. It is comforting that the exotica of cosmic strings do seem to be a natural consequence[47] of the current, in vogue, superstring Theories of Everything (T.O.E.).

Acknowledgements to co-workers J. Charlton, K. Olive, A. Melott, G. Steigman, A. Szalay, and M. Turner are gratefully given. I also acknowledge many useful discussions with N. Turok and P.J.E. Peebles. This work was supported in part by NSF AST 85-15447, and by DOE DE-FG02-85ER40234 at the University of Chicago, and was prepared at the Aspen Center for Physics.

Table I: Models and Problems

	Hot & Adiabatic	Cold & Adiabatic	Cold & Strings	Hot & Strings	Cold & Explosions	Hot & Explosions
$\Omega = 1$ with $\Omega_{cluster} \sim 0.2$	o.k.	requires ad hoc biasing	requires large cluster–galaxy correlation	requires large cluster–galaxy correlation	requires special biasing	o.k.
Large Cluster–Cluster correlation function	difficult	no	o.k.	o.k. if not destroyed by pancaking	requires fine tuning	requires fine tuning
Filaments, sheets, and voids structure at ~ 40Mpc	o.k.	difficult	not easy	maybe	difficult	o.k.
Large scale high velocities	o.k.	no, worse with biasing	no	maybe	difficult	o.k.
Galaxy formation by $z \gtrsim 4$	no	o.k.	o.k.	marginal	o.k.	depends on seeds
Galaxy mass spectrum	pancake fragmentation	depends on biasing scheme	maybe	probably o.k.	maybe	maybe
Galaxy angular momentum	pancake fragmentation	o.k.	probably o.k.	probably o.k.	probably o.k.	probably o.k.
Contents of voids	mostly hot stuff	~ 90% cold ~ 10% baryons	~ 90% cold ~ 10% baryons	$\gtrsim$ 90% hot $\lesssim$ 10% baryons	~ 90% cold ~ 10% baryons	mostly hot stuff
Halos of galaxies (including dwarfs)	mostly baryonic	~ 90% cold ~ 10% baryonic	~ 90% cold ~ 10% baryonic	> 10% baryonic < 90% hot	~ 90% cold ~ 10% baryonic	mostly baryonic

12. References

1. deLapparant, V., Geller, M. and Huchra, J. 1986, Center for Astrophysics preprint
2. Koo, D. and Kron, R. 1986, in preparation.
3. Kirschner, R., Oemler, G., Schecter, P., and Shectman, S. 1982, *Ap.J.* **248**, L57.
4. Faber, S., Aaronson, M., Lynden-Bell, D. 1986, *Proc. of Hawaii Symposium on Large-Scale Structure.*
5. Bahcall, N. and Soniera, R. 1983, *Ap.J.* **270**, 20; Klypin and Khlopov 1983, *Soviet Astron. Lett.* **9**, 41.
6. Szalay, A. and Schramm, D. 1985, *Nature* **314**, 718.
7. Gott, J.R., Gunn, J., Schramm, D.N., and Tinsley, B.M. 1974, *Ap.J.* **194**, 543.
8. Freese, K. and Schramm, D. 1984, *Nucl. Physics* **B233**, 167.
9. Yang, J., Turner, M., Steigman, G., Schramm, D., and Olive, K. 1984 *Ap.J.* **281**, 493.
10. Steigman, G., Schramm, D.N., and Gunn, J.E. 1977, *Phys.Lett.* **B66**, 502.
11. Cline, D. 1986, The 6th Proton–Anti-proton Conference, Aachen, West Germany, review talk.
12. Vittorio, N. and Silk, J. 1984, *Ap.J.* **L39**.
13. Bond, J., Efstathiou, G., and Silk, J. 1980, *Phys.Rev.Lett.* **45**, 1980.
14. Rowan-Robinson, M. 1986, in The Proc. ESO/CERN Symposium on Cosmology.
15. Loh, E. and Spillar, E. 1986, Princeton University preprint
16. Ostriker, J. and Cowie, L. 1980, *Ap.J.* **243**, L127.
17. Olive, K. and Schramm, D.N. 1986 *Comments on Nuclear and Particle Physics*, in press.
18. Schramm, D. and Steigman, G. 1981, *Ap.J.* **243**, 1.
19. Lubimov, A. 1986, in this volume.
20. Frenk, C., White, S., and Davis, M. 1983, *Ap.J.* **271**, 417.
21. Davis, M. 1986 *Proc. 1984 Inner Space/Outer Space*, University of Chicago Press.
22. Melott, A. 1986 *Proc. 1984 Inner Space/Outer Space*, University of Chicago Press.
23. Blumenthal, G., Faber, S., Primack, J., and Rees, M. 1984, *Nature* **311**, 517.
24. Melott, A., Einasto, J., Saar, E., Suisalu, I., Klypin, A., and Shandarin, S. 1983, *Phys.Rev.Lett* **51**, 935.
25. Efstathiou, G., Frenk, C., White, S., and Davis, M. 1985 *Ap.J.Suppl.* **57**, 241.
26. Schramm, D. 1985, *Proc. 1984 Rome Conf. on Microwave Background.*
27. Bardeen, J., Bond, J., Kaiser, N., and Szalay, A. 1985, submitted to *Ap.J.*.
28. Turok, N. 1985, U.C. Santa Barbara preprint
29. Melott, A. 1986, Univ. of Chicago preprint.
30. Charlton, J. and Schramm, D. 1986, submitted to *Ap.J.*.
31. Vilenkin, A. 1985, *Physics Reports* **121**, 1.
32. Vittorio, N. and Schramm, D. 1985, *Comments on Nuclear and Particle Physics* **15**, 1.
33. Turok, D.N. and Schramm, D.N. 1986, in preparation
34. Bennet, D. 1986, SLAC preprint
35. Vilenkin, A. 1986, Tufts Univesity preprint
36. Rashiputi 1986, preprint
37. Pagels, H. 1986. Rockefeller University preprint.
38. Peebles, P.J.E. 1981, *The Large Scale Structure of the Universe*, Princeton University Press.
39. Olive, K. and Seckel, D. 1986, FNAL preprint.
40. Crawford, M. and Schramm, D. 1982, *Nature* **298**, 538.
41. Witten, E. 1984, *Phys.Rev.* **D30**, 272.
42. Applegate, J. and Hogan, C. 1985, *Phys.Rev.* **D31**, 3037.
43. Alcock, C. and Farhi, J. 1985, MIT preprint.
44. Freese, K., Price, R., and Schramm, D. 1983, *Ap.J.* **275**, 405.
45. Vishniac, E., Ostriker, J., and Bertschinger, E. 1985, Princeton University preprint.
46. Turok, N. and Schramm, D. 1984, *Nature* **312**, 598.
47. Witten, E. 1985, *Physics Letters* **B153**, 243.

COLD DARK MATTER CANDIDATES

Mark Srednicki*
Department of Physics
University of California
Santa Barbara, California 93106

ABSTRACT

The possibility of detection of indirect signatures of plausible cold dark matter candidates is discussed.

The evidence for a universe dominated by cold dark matter is well known. In this talk I will discuss two particular issues: what constitutes a plausible cold dark matter candidate, and possibilities for indirect detection of some of these candidates.

Particles constituting most of the mass of the universe must be stable on (at least) the time scale set by the age of the universe. There are only three particles we know of which are absolutely stable: the electron, the photon, and the lightest spin one-half particle (which may or may not be one of the three presently known neutrinos). Their stability is guaranteed by, respectively, conservation of electric charge, electromagnetic gauge invariance, and conservation of angular momentum. These are deeply held principles of physics. To the list of known particles stable on cosmological time scales we may add the proton. The proton is almost stable because it is not possible to write down a renormalizable interaction among the particles of the Standard Model that would allow it to decay. That exhausts the list of known long-lived particles. All others decay into these four (and their antiparticles) at a rate which is extremely fast from the universe's point of view.

This means that we ought to be careful about postulating new stable particles. We ought to have very good reasons, comparable to conservation of angular momentum or gauge invariance, for suggesting their existence. I know of only two such candidate principles: time reversal invariance of the strong interactions, and supersymmetry. The first leads to the axion,[1] a very light ($\sim 10^{-5}$ eV) spin zero particle which has a lifetime against decay into two photons much longer than the age of the universe. A nonthermal process[2] during a very early phase transition can fill up the universe with axions which are at rest, and we have instant cold dark matter. Supersymmetry by itself does not lead to a new stable particle.

* Supported in part by the National Science Foundation under Grant No. PHY83-13324, and by the Alfred P. Sloan Foundation.

It must be combined with a less noble principle, conservation of R-parity.[3] But supersymmetry without R-parity leads to disasterous phenomenology (the proton lives for less than a second), so if we believe in supersymmetry manifest at ordinary energies, we are forced to believe in R-parity as well. Then the lightest supersymmetric partner of ordinary particles becomes at least as stable as the proton. Any other cold dark matter candidates, such as heavy stable neutrinos, are cut from whole cloth. No plausible principle leads to their stability.

I will not have anything more to say about axions. Of the supersymmetric candidates, the most likely[4] are colorless and electrically neutral: the photino, the higgsino, and the three scalar neutrinos. These particles would have annihilation and nucleon scattering cross sections of roughly weak interaction strength. This means that a galactic halo made out of them would not undergo viscous collapse to a disk, which is of course what we want.

If our halo is made of these particles, how can we find out? It may be possible to detect them directly.[5] This is a tremendously exciting possibility, but one which I will not discuss. Another idea is to look for telltale astrophysical signatures. It is this possibility to which I now turn.

Halo particles are moving at about $10^{-3}c$, so their total energy is in their mass. If two of them collide and annihilate, they will turn into some ordinary particles with total energy equal to twice the mass of the dark matter particle. If this mass is a few GeV, and antiprotons are among the allowed annihilation products, we have a source of low energy, cosmic ray antiprotons.[6] We can estimate the flux because we know the annihilation cross section. We know the annihilation cross section because it determines the relic mass density of the particles (which we want to be the critical density).[7,4] The result[8] is that photinos weighing about 3 GeV could explain the anomalously high flux of low energy cosmic ray antiprotons which has been observed.[6] Higgsinos and sneutrinos do not produce enough antiprotons.[9] It has been suggested that heavier photinos can account for observed cosmic ray antiprotons at all energies.[10] However, this requires scalar quark and lepton masses below experimental limits.[11] (These masses enter because they control the annihilation cross section, which must be larger in order to make *all* the cosmic ray antiprotons.) It also requires a mean photino mass density well below critical, and a local photino mass density well above most estimates.[12]

Annihilations of photinos or higgsinos into a quarkonium state plus a monochromatic photon leads to sharp peaks in the diffuse gamma ray background.[13] Unfortunately, the branching ratio for this process is probably too small[14] to make the peaks visible in the forseeable future.

This process, as well as that of gamma production in an ordinary annihilation, are significantly enhanced if the galactic bulge contains a large percentage of dark matter.[15] Constraints on the amount of various dark matter candidates in the bulge can be set.

Dark matter particles can scatter off nuclei in astrophysical bodies, lose energy, and become gravitationally bound.[16] Once trapped inside, these particles can annihilate. Neutrinos produced in these annihilations will escape.[17] If they have sufficient energy, they can be seen by proton decay detectors. However, photinos or higgsinos only interact with nuclear spins,[5] and thus pass through planets, as planets are composed mostly of spinless nuclei.[18,19] The sun, though, is made mostly of spin one-half protons, and traps[16,20] $7 \times 10^{28} m^{-1} \sigma_{p\,36}$ photinos or higgsinos per second for nominal values of astrophysical parameters (such as the galactic rotation velocity). Here $\sigma_{p\,36}$ is the cross section for scattering off protons in units of 10^{36} cm^2, and m is particle's mass in GeV. The sun and planets trap scalar neutrinos, whose cross section for scattering off nuclei grows as the square of the number of neutrons, regardless of spin.[5] The trapping rate for the sun is the same, except that $\sigma_{p\,36}$ is replaced by a suitable average over solar composition.[20] For the earth, iron is the most important, and the rate is[18,19] $5 \times 10^{21}(1 + m^2/m_{\mathrm{Fe}}^2)^{-1}\sigma_{\mathrm{Fe}\,36}$ particles per second. Here $\sigma_{\mathrm{Fe}\,36}$ is the scattering cross section off iron in units of 10^{36} cm^2, and m_{Fe} is the mass of an iron nucleus. The dependence on the dark matter particle's mass is different from the case of solar trapping because the earth captures only the low velocity tail of the thermal distribution of halo particles.[16] Once a steady state is achieved, each trapped particle must be balanced by a particle disappearing by either evaporation or annihilation. Annihilation will dominate for heavier particles, as they tend to sink to the center. Estimates of the particle mass at which annihilation begins to dominate range from 3.3 GeV[21] to 6 GeV.[18] The resulting fluxes of neutrinos are roughly comparable to atmospheric background.[17−20,22] Limits on the flux of muon neutrinos and antineutrinos from the sun, with energies in various bins, have recently been set by the IMB collaboration.[23] These limits can be compared with theoretical predictions of the differential flux dF/dE.[24] It turns out[25] that scalar muon neutrinos can be ruled out as dark matter, but no useful limits can be set on photinos or higgsinos as dark matter.

REFERENCES

[1] R. D. Peccei and H. R. Quinn, *Phys. Rev. Lett.* **38** (1977) 1440; *Phys. Rev. D* **16** (1977) 1791;

S. Weinberg, *Phys. Rev. Lett.* **40** (1978) 223;

F. Wilczek, *Phys. Rev. Lett.* **40** (1978) 279;

J. E. Kim, *Phys. Rev. Lett.* **43** (1979) 103;

M. A. Shifman, A. I. Vainshtein, and V. I. Zakharov, *Nucl. Phys.* **B166** (1980) 493;

M. Dine, W. Fischler, and M. Srednicki, *Phys. Lett.* **104B** (1981) 199.

[2] J. Preskill, M. B. Wise, and F. Wilczek, *Phys. Lett.* **120B** (1983) 127;

L. Abbott and P. Sikivie, *Phys. Lett.* **120B** (1983) 133;

M. Dine and W. Fischler, *Phys. Lett.* **120B** (1983) 137.

[3] For a review see H. P. Nilles, *Phys. Rep.* **110** (1984) 1.

[4] J. Ellis, J. S. Hagelin, D. V. Nanopoulos, K. Olive, and M. Srednicki, *Nucl. Phys.* **B238** (1984) 453.

[5] M. Goodman and E. Witten, *Phys. Rev.* D **31** (1984) 3059.

[6] J. Silk and M. Srednicki, *Phys. Rev. Lett.* **53** (1984) 624.

[7] B. Lee and S. Weinberg, *Phys. Rev. Lett.* **39** (1977) 165;

H. Goldberg, *Phys. Rev. Lett.* **50** (1983) 1419.

[8] A. Buffington, S. Schindler, and C. Pennypacker, *Astrophys. J.* **248** (1981) 1179.

[9] J. Hagelin and G. L. Kane, *Nucl. Phys.* **B263** (1986) 399.

[10] S. Rudaz, T. Walsh, and F. Stecker, *Phys. Rev. Lett.* **55** (1985) 2622.

[11] R. M. Barnett, H. E. Haber, and G. L. Kane, *Nucl. Phys.* **B267** (1986) 625.

[12] J. Bahcall and R. Soneira, *Astrophys. J. Supp.* **44** (1980) 73.

[13] M. Srednicki, S. Theisen, and J. Silk, *Phys. Rev. Lett.* **56** (1986) 263; 1883(E).

[14] S. Rudaz, *Phys. Rev. Lett.* **56** (1986) 2128.

[15] J. Silk and H. Bloemen, Berkeley preprint (1986).

[16] W. H. Press and D. N. Spergel, *Astrophys. J.* **296** (1985) 679.

[17] J. Silk, K. Olive, and M. Srednicki, *Phys. Rev. Lett.* **55** (1985) 257.

[18] L. M. Krauss, M. Srednicki, and F. Wilczek, *Phys. Rev.* D **33** (1986) 2079.

[19] K. Freese, *Phys. Lett.* **167B** (1986) 295.

[20] M. Srednicki, K. Olive, and J. Silk, *Nucl. Phys.* **B279** (1987) 804.

[21] K. Griest and D. Seckel, CERN preprint (1986).

[22] T. Gaisser, G. Steigman, and S. Tilav, *Phys. Rev.* D **34** (1986) 2206.

[23] J. M. LoSecco *et al.*, Notre Dame preprint (1986).

[24] J. S. Hagelin, K. W. Ng, and K. A. Olive, Minnesota preprint (1986).

[25] K. W. Ng, K. A. Olive, and M. Srednicki, in preparation.

WORKSHOP PARTICIPANTS

David Bennett

Stanford Linear Accelerator Center
Stanford, CA 94305

Pierre Binétruy

Laboratoire de Physique des Particles
Annecy-le-Vieux, France

J. Richard Bond

Canadian Institute for Theoretical Astrophysics
University of Toronto
Toronto, Ontario, Canada

Robert H. Brandenberger

DAMTP, University of Cambridge
Cambridge, England

Robert Cahn

Lawrence Berkeley Laboratory
Berkeley, CA 94720

Marc Davis

Department of Astronomy, University of California
Berkeley, CA 94720

Avishai Dekel

Racah Institute of Physics, The Hebrew University of Jerusalem
Jerusalem, Israel

John Faulkner

Lick Observatory, University of California at Santa Cruz
Santa Cruz, CA 95064

Ricardo A. Flores

Physics Department, Brandeis University
Waltham, MA 02254

Katherine Freese

Institute for Theoretical Physics
University of California at Santa Barbara
Santa Barbara, CA 93106

Carlos Frenk

Physics Department, University of Durham
Durham, England

Masataka Fukugita

Research Institute for Fundamental Physics, Kyoto University
Kyoto, Japan

Mary K. Gaillard

Lawrence Berkeley Laboratory and University of California
Berkeley, CA 94720

Belen Gavela

Theory Group, CERN
Geneva, Switzerland

Graciela Gelmini

Physics Department, Harvard University
Cambridge, MA 02138

Benjamin Grinstein

California Institute of Technology
Pasadena, CA 91125

Howard Haber

Department of Physics, University of California at Santa Cruz
Santa Cruz, CA 95064

Lawrence Hall

Lawrence Berkeley Laboratory and University of California
Berkeley, CA 94720

Ian Hinchliffe

Lawrence Berkeley Laboratory
Berkeley, CA 94720

Yehuda Hoffman

Physics Department, University of Pennsylvania
Philadelphia, PA 19104

Craig Hogan

Steward Observatory, University of Arizona
Tucson, AZ 85721

Roman Juszkiewicz

Department of Astronomy, University of California
Berkeley, CA 94720

Chung W. Kim

Department of Physics, Johns Hopkins University
Baltimore, MD 21218

Hideo Kodama

Department of Physics, University of Tokyo
Tokyo, Japan

Samir Mallik

Saha Institute of Nuclear Physics
Calcutta, India

Antonio Masiero

New York University
New York, NY 10003

Franco Occhionero

Institute of Astronomy, University of Rome
Rome, Italy

Keith Olive

Physics Department, University of Minnesota
Minneapolis, MN 55455

Stephen Parke

Theoretical Physics Department, Fermilab
Batavia, IL 60510

So-Young Pi

> Physics Department, Boston University
> Boston, MA 02215

Joel Primack

> Department of Physics, University of California at Santa Cruz
> Santa Cruz, CA 95064

Graham Ross

> Theoretical Physics Department, Oxford University
> Oxford, England

Pierre Salati

> Laboratoire de Physique des Particules
> Annecy-le-Vieux, France

Jose L. Sanz

> Department of Theoretical Physics, University of Santander
> Santander, Spain

David Schramm

> Astronomy and Astrophysics Center, University of Chicago
> Chicago, IL 60637

Ronald Shellard

> Pontifícia Universidade Católica
> Rio de Janeiro, Brasil

Marc Sher

> Physics Department, University of California at Santa Cruz
> Santa Cruz, CA 95064

Joseph Silk

> Department of Astronomy, University of California
> Berkeley, CA 94720

David Spergel

> Institute for Advanced Study
> Princeton, NJ 08540

Mark Srednicki

Department of Physics, University of California at Santa Barbara
Santa Barbara, CA 93106

Leo Stodolsky

Max-Planck Institute
Munich, Germany

Alexander Szalay

Department of Astronomy, Eotvos University
Budapest, Hungary

Katsumi Tanaka

Physics Department, Ohio State University
Columbus, OH 43210

Jay Villumsen

California Institute of Technology
Pasadena, CA 91125

Ethan Vishniac

Department of Astronomy, University of Texas
Austin, TX 78712

Nicola Vittorio

Institute of Astronomy, University of Rome
Rome, Italy

Thomas Weiler

Department of Physics and Astronomy, Vanderbilt University
Nashville, TN 37235

Simon White

Steward Observatory, University of Arizona
Tucson, AZ 85721

To Barb
from Dear Friend
Diane in
Oakhurst.
June 10, 2002